GALOIS THEORY

EMIL ARTIN

Edited and supplemented with a
section on Applications by
ARTHUR N. MILGRAM

DOVER PUBLICATIONS, INC.
Mineola, New York

Bibliographical Note

This Dover edition, first published in 1998, is an unabridged and unaltered republication of the last corrected printing of the 1944 second, revised edition of the work first published by The University of Notre Dame Press in 1942 as Number 2 in the series, *Notre Dame Mathematical Lectures.*

Library of Congress Cataloging-in-Publication Data

Artin, Emil, 1898–1962.
Galois theory / by Emil Artin ; edited and supplemented with a selection on applications by Arthur N. Milgram.
p. cm.
"An unabridged and unaltered republication of the last corrected printing of the 1944 second, revised edition of the work first published by the University of Notre Dame Press in 1942 as number 2 in the series, Notre Dame mathematical lectures"—T.p. verso
Includes bibliographical references (p. –) and index.
ISBN 0-486-62342-4 (pbk.)
1. Galois theory. I. Milgram, Arthur N. (Arthur Norton), 1912– . II. Title.
QA214.A77 1998
512'.3—dc21 97–51372
CIP

Manufactured in the United States of America
Dover Publications, Inc., 31 East 2nd Street, Mineola, N. Y. 11501

TABLE OF CONTENTS

(The sections marked with an asterisk have been herein added to the content of the first edition)

I LINEAR ALGEBRA

A. Fields.

A field is a set of elements in which a pair of operations called multiplication and addition is defined analogous to the operations of multiplication and addition in the real number system (which is itself an example of a field). In each field F there exist unique elements called o and 1 which, under the operations of addition and multiplication, behave with respect to all the other elements of F exactly as their correspondents in the real number system. In two respects, the analogy is not complete: 1) multiplication is not assumed to be commutative in every field, and 2) a field may have only a finite number of elements.

More exactly, a field is a set of elements which, under the above mentioned operation of addition, forms an additive abelian group and for which the elements, exclusive of zero, form a multiplicative group and, finally, in which the two group operations are connected by the distributive law. Furthermore, the product of o and any element is defined to be o.

If multiplication in the field is commutative, then the field is called a commutative field.

B. Vector Spaces.

If V is an additive abelian group with elements A, B, . . . , F a field with elements a, b, . . . , and if for each $a \in F$ and $A \in V$

the product aA denotes an element of V, then V is called a (left) vector space over F if the following assumptions hold:

1) $a(A + B) = aA + aB$

2) $(a + b)A = aA + bA$

3) $a(bA) = (ab)A$

4) $1A = A$

The reader may readily verify that if V is a vector space over F, then $oA = O$ and $aO = O$ where o is the zero element of F and O that of V. For example, the first relation follows from the equations:

$$aA = (a + o)A = aA + oA$$

Sometimes products between elements of F and V are written in the form Aa in which case V is called a right vector space over F to distinguish it from the previous case where multiplication by field elements is from the left. If, in the discussion, left and right vector spaces do not occur simultaneously, we shall simply use the term "vector space."

C. Homogeneous Linear Equations.

If in a field F, a_{ij}, $i = 1, 2, \ldots, m$, $j = 1, 2, \ldots, n$ are $m \cdot n$ elements, it is frequently necessary to know conditions guaranteeing the existence of elements in F such that the following equations are satisfied:

$$(1) \qquad \begin{array}{c} a_{11}x_1 + a_{12}x_2 + \ldots + a_{1n}x_n = o. \\ \vdots \\ a_{m1}x_1 + a_{m2}x_2 + \ldots + a_{mn}x_n = o. \end{array}$$

The reader will recall that such equations are called linear homogeneous equations, and a set of elements, $x_1, x_2, \ldots, x_n$ of F, for which all the above equations are true, is called a solution of the system. If not all of the elements $x_1, x_2, \ldots, x_n$ are o the solution is called non-trivial; otherwise, it is called trivial.

THEOREM 1. A system of linear homogeneous equations always has a non-trivial solution if the number of unknowns exceeds the number of equations.

The proof of this follows the method familiar to most high school students, namely, successive elimination of unknowns. If no equations in $n > 0$ variables are prescribed, then our unknowns are unrestricted and we may set them all $= 1$.

We shall proceed by complete induction. Let us suppose that each system of k equations in more than k unknowns has a non-trivial solution when $k < m$. In the system of equations (1) we assume that $n > m$, and denote the expression $a_{i1}x_1 + \ldots + a_{in}x_n$ by L_i, $i = 1, 2, \ldots, m$. We seek elements $x_1, \ldots, x_n$ not all o such that $L_1 = L_2 = \ldots = L_m = o$. If $a_{ij} = o$ for each i and j, then any choice of $x_1, \ldots, x_n$ will serve as a solution. If not all a_{ij} are o, then we may assume that $a_{11} \neq o$, for the order in which the equations are written or in which the unknowns are numbered has no influence on the existence or non-existence of a simultaneous solution. We can find a non-trivial solution to our given system of equations, if and only if we can find a non-trivial solution to the following system:

$$
\begin{aligned}
L_1 &= o \\
L_2 - a_{21}a_{11}^{-1}L_1 &= o \\
\cdot \quad \cdot \quad \cdot \quad \cdot \quad \cdot \quad & \quad \cdot \\
L_m - a_{m1}a_{11}^{-1}L_1 &= o
\end{aligned}
$$

For, if $x_1, \ldots, x_n$ is a solution of these latter equations then, since $L_1 = o$, the second term in each of the remaining equations is o and, hence, $L_2 = L_3 = \ldots = L_m = o$. Conversely, if (1) is satisfied, then the new system is clearly satisfied. The reader will notice that the new system was set up in such a way as to "eliminate" x_1 from the last m-1 equations. Furthermore, if a non-trivial solution of the last m-1 equations, when viewed as equations in $x_2, \ldots, x_n$, exists then taking $x_1 = - a_{11}^{-1}(a_{12}x_2 + a_{13}x_3 + \ldots + a_{1n}x_n)$ would give us a solution to the whole system. However, the last m-1 equations have a solution by our inductive assumption, from which the theorem follows.

Remark: If the linear homogeneous equations had been written in the form $\Sigma x_j a_{ij} = o$, $j = 1, 2, \ldots, n$, the above theorem would still hold and with the same proof although with the order in which terms are written changed in a few instances.

D. Dependence and Independence of Vectors.

In a vector space V over a field F, the vectors $A_1, \ldots, A_n$ are called dependent if there exist elements $x_1, \ldots, x_n$, not all o, of F such that $x_1A_1 + x_2A_2 + \ldots + x_nA_n = O$. If the vectors $A_1, \ldots, A_n$ are not dependent, they are called independent.

The dimension of a vector space V over a field F is the maximum number of independent elements in V. Thus, the dimension of V is n if there are n independent elements in V, but no set of more than n independent elements.

A system $A_1, \ldots, A_m$ of elements in V is called a generating system of V if each element A of V can be expressed linearly in terms of $A_1, \ldots, A_m$, i.e., $A = \sum_{i=1}^{m} a_i A_i$ for a suitable choice of a_i, $i = 1, \ldots, m$, in F.

THEOREM 2. In any generating system the maximum number of independent vectors is equal to the dimension of the vector space.

Let $A_1, \ldots, A_m$ be a generating system of a vector space V of dimension n. Let r be the maximum number of independent elements in the generating system. By a suitable reordering of the generators we may assume $A_1, \ldots, A_r$ independent. By the definition of dimension it follows that $r \leq n$. For each j, $A_1, \ldots, A_r$. A_{r+j} are dependent, and in the relation

$$a_1 A_1 + a_2 A_2 + \ldots + a_r A_r + a_{r+j} A_{r+j} = O$$

expressing this, $a_{r+j} \neq o$, for the contrary would assert the dependence of $A_1, \ldots, A_r$. Thus,

$$A_{r+j} = - a_{r+j}^{-1} [a_1 A_1 + a_2 A_2 + \ldots + a_r A_r].$$

It follows that $A_1, \ldots, A_r$ is also a generating system since in the linear relation for any element of V the terms involving A_{r+j}, $j \neq o$, can all be replaced by linear expressions in $A_1, \ldots, A_r$.

Now, let $B_1, \ldots, B_t$ be any system of vectors in V where $t > r$, then there exist a_{ij} such that $B_j = \sum_{i=1}^{r} a_{ij} A_i$, $j = 1, 2, \ldots, t$, since the A_i's form a generating system. If we can show that $B_1, \ldots, B_t$ are dependent, this will give us $r \geq n$, and the theorem will follow from this together with the previous inequality $r \leq n$. Thus, we must exhibit the existence of a non-trivial solution out of F of the equation

$$x_1 B_1 + x_2 B_2 + \ldots + x_t B_t = O.$$

To this end, it will be sufficient to choose the x_i's so as to satisfy the linear equations $\sum_{j=1}^{t} x_j a_{ij} = o$, $i = 1, 2, \ldots, r$, since these expressions will be the coefficients of A_i when in $\sum_{j=1}^{t} x_j B_j$ the B_j's are replaced by $\sum_{i=1}^{r} a_{ij} A_i$ and terms are collected. A solution to the equations $\sum_{j=1}^{t} x_j a_{ij} = o$, $i = 1, 2, \ldots, r$, always exists by Theorem 1.

Remark: Any n independent vectors $A_1, \ldots, A_n$ in an n dimensional vector space form a generating system. For any vector A, the vectors $A, A_1, \ldots, A_n$ are dependent and the coefficient of A, in the dependence relation, cannot be zero. Solving for A in terms of $A_1, \ldots, A_n$, exhibits $A_1, \ldots, A_n$ as a generating system.

A subset of a vector space is called a subspace if it is a subgroup of the vector space and if, in addition, the multiplication of any element in the subset by any element of the field is also in the subset. If $A_1, \ldots, A_s$ are elements of a vector space V, then the set of all elements of the form $a_1 A_1 + \ldots + a_s A_s$ clearly forms a subspace of V. It is also evident, from the definition of dimension, that the dimension of any subspace never exceeds the dimension of the whole vector space.

An s-tuple of elements $(a_1, \ldots, a_s)$ in a field F will be called a row vector. The totality of such s-tuples form a vector space if we define

α) $(a_1, a_2, \ldots, a_s) = (b_1, b_2, \ldots, b_s)$ if and only if $a_1 = b_1$, $i = 1, \ldots, s$,

β) $(a_1, a_2, \ldots, a_s) + (b_1, b_2, \ldots, b_s) = (a_1 + b_1, a_2 + b_2, \ldots, a_s + b_s)$,

γ) $b(a_1, a_2, \ldots, a_s) = (ba_1, ba_2, \ldots, ba_s)$, for b an element of F.

When the s-tuples are written vertically, $\begin{pmatrix} a_1 \\ \cdot \\ \cdot \\ \cdot \\ a_s \end{pmatrix}$

they will be called column vectors.

THEOREM 3. The row (column) vector space F^n of all n-tuples from a field F is a vector space of dimension n over F.

The n elements

$$\begin{aligned} \epsilon_1 &= (1, o, o, \ldots, o) \\ \epsilon_2 &= (o, 1, o, \ldots, o) \\ &\cdot \\ &\cdot \\ &\cdot \\ \epsilon_n &= (o, o, \ldots, o, 1) \end{aligned}$$

are independent and generate F^n. Both remarks follow from the relation $(a_1, a_2, \ldots, a_n) = \Sigma a_i \epsilon_i$.

We call a rectangular array

$$\begin{pmatrix} a_{11} a_{12} \ldots a_{1n} \\ a_{21} a_{22} \ldots a_{2n} \\ \cdot \quad \cdot \qquad \cdot \\ \cdot \quad \cdot \qquad \cdot \\ \cdot \quad \cdot \qquad \cdot \\ a_{m1} a_{m2} \ldots a_{mn} \end{pmatrix}$$

of elements of a field F a matrix. By the right row rank of a matrix, we mean the maximum number of independent row vectors among the rows $(a_{i1}, \ldots, a_{in})$ of the matrix when multiplication by field elements is from the right. Similarly, we define left row rank, right column rank and left column rank.

THEOREM 4. In any matrix the right column rank equals the left row rank and the left column rank equals the right row rank. If the field is commutative, these four numbers are equal to each other and are called the rank of the matrix.

Call the column vectors of the matrix $C_1, \ldots, C_n$ and the row vectors $R_1, \ldots, R_m$. The column vector O is $\begin{pmatrix} o \\ o \\ \cdot \\ \cdot \\ o \end{pmatrix}$ and any dependence $C_1x_1 + C_2x_2 + \ldots + C_nx_n = O$ is equivalent to a solution of the equations

$$
(1) \quad \begin{matrix} a_{11}x_1 + a_{12}x_2 + \ldots + a_{1n}x_n = o \\ \cdot \qquad\qquad\qquad \cdot \\ \cdot \qquad\qquad\qquad \cdot \\ \cdot \qquad\qquad\qquad \cdot \\ a_{m1}x_1 + a_{m2}x_2 + \ldots + a_{mn}x_n = o. \end{matrix}
$$

Any change in the order in which the rows of the matrix are written gives rise to the same system of equations and, hence, does not change the column rank of the matrix, but also does not change the row rank since the changed matrix would have the same set of row vectors. Call c the right column rank and r the left row rank of the matrix. By the above remarks we may assume that the first r rows are independent row vectors. The row vector space generated by all the rows of the matrix has, by Theorem 1, the dimension r and is even generated by the first r rows. Thus, each row after the rth is linearly expressible in terms of the first r rows. Consequently, any solution of the first r equations in (1) will be a solution of the entire system since any of the last n-r equations is obtainable as a linear combination of the first r. Conversely, any solution of (1) will also be a solution of the first r equations. This means that the matrix

$$\begin{pmatrix} a_{11} a_{12} \dots a_{1n} \\ \cdot \quad \cdot \quad \cdot \\ \cdot \quad \cdot \quad \cdot \\ \cdot \quad \cdot \quad \cdot \\ a_{r1}\ a_{r2} \dots a_{rn} \end{pmatrix}$$

consisting of the first r rows of the original matrix has the same right column rank as the original. It has also the same left row rank since the r rows were chosen independent. But the column rank of the amputated matrix cannot exceed r by Theorem 3. Hence, $c \leq r$. Similarly, calling c' the left column rank and r' the right row rank, $c' \leq r'$. If we form the transpose of the original matrix, that is, replace rows by columns and columns by rows, then the left row rank of the transposed matrix equals the left column rank of the original. If then to the transposed matrix we apply the above considerations we arrive at $r \leq c$ and $r' \leq c'$.

E. Non-homogeneous Linear Equations.

The system of non-homogeneous linear equations

$$\begin{array}{lllll} a_{11}x_1 & + a_{12}x_2 + \dots & + & a_{1n}x_n & = b_1 \\ a_{21}x_1 & + \dots\dots\dots\dots & + & a_{2n}x_n & = b_2 \\ \cdot & & & \cdot & \cdot \\ \cdot & & & \cdot & \cdot \\ \cdot & & & \cdot & \cdot \\ a_{m1}x_1 & + \dots\dots\dots\dots & + & a_{mn}x_m & = b_m \end{array} \tag{2}$$

has a solution if and only if the column vector $\begin{pmatrix} b_1 \\ \cdot \\ \cdot \\ \cdot \\ b_m \end{pmatrix}$ lies

in the space generated by the vectors $\begin{pmatrix} a_{11} \\ \cdot \\ \cdot \\ \cdot \\ a_{m1} \end{pmatrix}, \dots, \begin{pmatrix} a_{1n} \\ \cdot \\ \cdot \\ \cdot \\ a_{mn} \end{pmatrix}$

This means that there is a solution if and only if the right column rank of the matrix $\begin{pmatrix} a_{11} \cdots a_{1n} \\ \cdot \quad\quad \cdot \\ \cdot \quad\quad \cdot \\ \cdot \quad\quad \cdot \\ a_{m1} \cdots a_{mn} \end{pmatrix}$ is the same as the right column rank of the augmented matrix $\begin{pmatrix} a_{11} \cdots a_{1n} b_1 \\ \cdot \quad\quad \cdot \quad \cdot \\ \cdot \quad\quad \cdot \quad \cdot \\ \cdot \quad\quad \cdot \quad \cdot \\ a_{m1} \cdots a_{mn} b_m \end{pmatrix}$ since the vector space generated by the original must be the same as the vector space generated by the augmented matrix and in either case the dimension is the same as the rank of the matrix by Theorem 2.

By Theorem 4, this means that the row ranks are equal. Conversely, if the row rank of the augmented matrix is the same as the row rank of the original matrix, the column ranks will be the same and the equations will have a solution.

If the equations (2) have a solution, then any relation among the rows of the original matrix subsists among the rows of the augmented matrix. For equations (2) this merely means that like combinations of equals are equal. Conversely, if each relation which subsists between the rows of the original matrix also subsists between the rows of the augmented matrix, then the row rank of the augmented matrix is the same as the row rank of the original matrix. In terms of the equations this means that there will exist a solution if and only if the equations are consistent, i.e., if and only if any dependence between the left hand sides of the equations also holds between the right sides.

THEOREM 5. If in equations (2) $m = n$, there exists a unique solution if and only if the corresponding homogeneous equations

$$\begin{matrix} a_{11}x_1 + a_{12}x_2 + \dots + a_{1n}x_n = 0 \\ \vdots \qquad\qquad\qquad \vdots \\ a_{n1}x_1 + a_{n2}x_2 + \dots + a_{nn}x_n = 0 \end{matrix}$$

have only the trivial solution.

If they have only the trivial solution, then the column vectors are independent. It follows that the original n equations in n unknowns will have a unique solution if they have any solution, since the difference, term by term, of two distinct solutions would be a non-trivial solution of the homogeneous equations. A solution would exist since the n independent column vectors form a generating system for the n-dimensional space of column vectors.

Conversely, let us suppose our equations have one and only one solution. In this case, the homogeneous equations added term by term to a solution of the original equations would yield a new solution to the original equations. Hence, the homogeneous equations have only the trivial solution.

F. Determinants.[1]

The theory of determinants that we shall develop in this chapter is not needed in Galois theory. The reader may, therefore, omit this section if he so desires.

We assume our field to be commutative and consider the square matrix

1) Of the preceding theory only Theorem 1, for homogeneous equations and the notion of linear dependence are assumed known.

$$(1)\qquad \begin{pmatrix} a_{11}\,a_{12}\cdots a_{1n} \\ a_{21}\,a_{22}\cdots a_{2n} \\ \cdots\cdots\cdots \\ a_{n1}\,a_{n2}\cdots a_{nn} \end{pmatrix}$$

of n rows and n columns. We shall define a certain function of this matrix whose value is an element of our field. The function will be called the determinant and will be denoted by

$$(2)\qquad \begin{vmatrix} a_{11}\,a_{12}\cdots a_{1n} \\ a_{21}\,a_{22}\cdots a_{2n} \\ \cdots\cdots\cdots \\ a_{n1}\,a_{n2}\cdots a_{nn} \end{vmatrix}$$

or by $D(A_1, A_2, \ldots A_n)$ if we wish to consider it as a function of the column vectors $A_1, A_2, \ldots A_n$ of (1). If we keep all the columns but A_k constant and consider the determinant as a function of A_k, then we write $D_k(A_k)$ and sometimes even only D.

<u>Definition.</u> A function of the column vectors is a determinant if it satisfies the following three axioms:

1. Viewed as a function of any column A_k it is linear and homogeneous, i.e.,

$$(3)\qquad D_k(A_k + A_k') = D_k(A_k) + D_k(A_k')$$

$$(4)\qquad D_k(cA_k) = c \cdot D_k(A_k)$$

2. Its value is $= 0$[1] if the adjacent columns A_k and A_{k+1} are equal.
3. Its value is $= 1$ if all A_k are the unit vectors U_k where

---

1) Henceforth, 0 will denote the zero element of a field.

$$(5)\quad U_1 = \begin{pmatrix} 1 \\ 0 \\ 0 \\ \cdot \\ \cdot \\ 0 \end{pmatrix};\ U_2 = \begin{pmatrix} 0 \\ 1 \\ \\ \cdot \\ \cdot \\ 0 \end{pmatrix} \cdots\cdots U_n = \begin{pmatrix} 0 \\ 0 \\ \\ \cdot \\ \cdot \\ 1 \end{pmatrix}$$

The question as to whether determinants exist will be left open for the present. But we derive consequences from the axioms:

a) If we put $c = 0$ in (4) we get: a determinant is 0 if one of the columns is 0.

b) $D_k(A_k) = D_k(A_k + cA_{k\pm 1})$ or a determinant remains unchanged if we add a multiple of one column to an adjacent column. Indeed

$$D_k(A_k + cA_{k\pm 1}) = D_k(A_k) + cD_k(A_{k\pm 1}) = D_k(A_k)$$

because of axiom 2.

c) Consider the two columns A_k and A_{k+1}. We may replace them by A_k and $A_{k+1} + A_k$; subtracting the second from the first we may replace them by $-A_{k+1}$ and $A_{k+1} + A_k$; adding the first to the second we now have $-A_{k+1}$ and A_k; finally, we factor out -1. We conclude: a determinant changes sign if we interchange two adjacent columns.

d) A determinant vanishes if any two of its columns are equal. Indeed, we may bring the two columns side by side after an interchange of adjacent columns and then use axiom 2. In the same way as in b) and c) we may now prove the more general rules:

e) Adding a multiple of one column to another does not change the value of the determinant.

f) Interchanging any two columns changes the sign of D.

g) Let $(\nu_1, \nu_2, \ldots \nu_n)$ be a permutation of the subscripts $(1, 2, \ldots n)$. If we rearrange the columns in $D(A_{\nu_1}, A_{\nu_2}, \ldots, A_{\nu_n})$ until they are back in the natural order, we see that

$$D(A_{\nu_1}, A_{\nu_2}, \ldots, A_{\nu_n}) = \pm D(A_1, A_2, \ldots, A_n).$$

Here $\pm$ is a definite sign that does not depend on the special values of the A_k. If we substitute U_k for A_k we see that $D(U_{\nu_1}, U_{\nu_2}, \ldots, U_{\nu_n}) = \pm 1$ and that the sign depends only on the permutation of the unit vectors.

Now we replace each vector A_k by the following linear combination A'_k of $A_1, A_2, \ldots, A_n$:

(6) $$A'_k = b_{1k}A_1 + b_{2k}A_2 + \ldots + b_{nk}A_n.$$

In computing $D(A'_1, A'_2, \ldots, A'_n)$ we first apply axiom 1 on A'_1 breaking up the determinant into a sum; then in each term we do the same with A'_2 and so on. We get

(7) $$D(A'_1, A'_2, \ldots, A'_n) = \sum_{\nu_1, \nu_2, \ldots, \nu_n} D(b_{\nu_1 1}A_{\nu_1}, b_{\nu_2 2}A_{\nu_2}, \ldots, b_{\nu_n n}A_{\nu_n})$$
$$= \sum_{\nu_1, \nu_2, \ldots, \nu_n} b_{\nu_1 1} \cdot b_{\nu_2 2} \cdot \ldots \cdot b_{\nu_n n} D(A_{\nu_1}, A_{\nu_2}, \ldots, A_{\nu_n})$$

where each ν_i runs independently from 1 to n. Should two of the indices ν_i be equal, then $D(A_{\nu_1}, A_{\nu_2}, \ldots, A_{\nu_n}) = 0$; we need therefore keep only those terms in which $(\nu_1, \nu_2, \ldots, \nu_n)$ is a permutation of $(1, 2, \ldots, n)$. This gives

(8) $$D(A'_1, A'_2, \ldots, A'_n)$$
$$= D(A_1, A_2, \ldots, A_n) \cdot \sum_{(\nu_1, \ldots, \nu_n)} \pm b_{\nu_1 1} \cdot b_{\nu_2 2} \cdot \ldots \cdot b_{\nu_n n}$$

where $(\nu_1, \nu_2, \ldots, \nu_n)$ runs through all the permutations of $(1, 2, \ldots, n)$ and where $\pm$ stands for the sign associated with that permutation. It is important to remark that we would have arrived at the same formula (8) if our function D satisfied only the first two of our axioms.

Many conclusions may be derived from (8).

We first assume axiom 3 and specialize the A_k to the unit vectors U_k of (5). This makes $A_k' = B_k$ where B_k is the column vector of the matrix of the b_{ik}. (8) yields now:

$$(9)\quad D(B_1, B_2, \ldots, B_n) = \sum_{(\nu_1, \nu_2, \ldots, \nu_n)} \pm\, b_{\nu_1 1} \cdot b_{\nu_2 2} \cdot \ldots \cdot b_{\nu_n n}$$

giving us an explicit formula for determinants and showing that they are uniquely determined by our axioms provided they exist at all.

With expression (9) we return to formula (8) and get

$$(10)\quad D(A_1', A_2', \ldots, A_n') = D(A_1, A_2, \ldots, A_n) D(B_1, B_2, \ldots, B_n).$$

This is the so-called multiplication theorem for determinants. At the left of (10) we have the determinant of an n-rowed matrix whose elements c_{ik} are given by

$$(11)\qquad c_{ik} = \sum_{\nu=1}^{n} a_{i\nu} b_{\nu k}.$$

c_{ik} is obtained by multiplying the elements of the i-th row of $D(A_1, A_2, \ldots, A_n)$ by those of the k-th column of $D(B_1, B_2, \ldots, B_n)$ and adding.

Let us now replace D in (8) by a function $F(A_1, \ldots, A_n)$ that satisfies only the first two axioms. Comparing with (9) we find

$$F(A_1', A_2', \ldots, A_n') = F(A_1, \ldots, A_n) D(B_1, B_2, \ldots, B_n).$$

Specializing A_k to the unit vectors U_k leads to

$$(12)\quad F(B_1, B_2, \ldots, B_n) = c \cdot D(B_1, B_2, \ldots, B_n)$$

with $c = F(U_1, U_2, \ldots, U_n)$.

Next we specialize (10) in the following way: If i is a certain subscript from 1 to n-1 we put $A_k = U_k$ for $k \neq i, i+1$ $A_i = U_i + U_{i+1}$, $A_{i+1} = 0$. Then $D(A_1, A_2, \ldots, A_n) = 0$ since one column is 0. Thus, $D(A_1', A_2', \ldots, A_n') = 0$; but this determinant differs from that of the elements b_{jk} only in the respect that the $i+1$-st row has been made equal to the i-th. We therefore see:

A determinant vanishes if two adjacent rows are equal.

Each term in (9) is a product where precisely one factor comes from a given row, say, the i-th. This shows that the determinant is linear and homogeneous if considered as function of this row. If, finally, we select for each row the corresponding unit vector, the determinant is $= 1$ since the matrix is the same as that in which the columns are unit vectors. This shows that a determinant satisfies our three axioms if we consider it as function of the row vectors. In view of the uniqueness it follows:

A determinant remains unchanged if we transpose the row vectors into column vectors, that is, if we rotate the matrix about its main diagonal.

A determinant vanishes if any two rows are equal. It changes sign if we interchange any two rows. It remains unchanged if we add a multiple of one row to another.

We shall now prove the existence of determinants. For a 1-rowed matrix a_{11} the element a_{11} itself is the determinant. Let us assume the existence of $(n-1)$-rowed determinants. If we consider the n-rowed matrix (1) we may associate with it certain $(n-1)$-rowed determinants in the following way: Let a_{ik} be a particular element in (1). We cancel the i-th row and k-th column in (1) and take the determinant of the remaining (n - 1)-rowed matrix. This determinant multiplied by $(-1)^{i+k}$ will be called the cofactor of a_{ik} and be denoted by A_{ik}. The distribution of the sign $(-1)^{i+k}$ follows the chessboard pattern, namely,

$$\begin{pmatrix} + & - & + & - & \cdot & \cdot & \cdot & \cdot \\ - & + & - & + & \cdot & \cdot & \cdot & \cdot \\ + & - & + & - & \cdot & \cdot & \cdot & \cdot \\ - & + & - & + & \cdot & \cdot & \cdot & \cdot \\ \cdot & \cdot & \cdot & \cdot & \cdot & \cdot & \cdot & \cdot \end{pmatrix}$$

Let i be any number from 1 to n. We consider the following function D of the matrix (1):

$$D = a_{i1}A_{i1} + a_{i2}A_{i2} + \dots + a_{in}A_{in}. \tag{13}$$

It is the sum of the products of the i-th row and their cofactors.

Consider this D in its dependence on a given column, say, A_k. For $\nu \neq k$, $A_{i\nu}$ depends linearly on A_k and $a_{i\nu}$ does not depend on it; for $\nu = k$, A_{ik} does not depend on A_k but a_{ik} is one element of this column. Thus, axiom 1 is satisfied. Assume next that two adjacent columns A_k and A_{k+1} are equal. For $\nu \neq k, k+1$ we have then two equal columns in $A_{i\nu}$ so that $A_{i\nu} = 0$. The determinants used in the computation of $A_{i,k}$ and $A_{i,k+1}$ are the same but the signs are opposite; hence, $A_{i,k} = -A_{i,k+1}$ whereas $a_{i,k} = a_{i,k+1}$. Thus $D = 0$ and axiom 2 holds. For the special case $A_\nu = U_\nu (\nu = 1, 2, \dots, n)$ we have $a_{i\nu} = 0$ for $\nu \neq i$ while $a_{ii} = 1$, $A_{ii} = 1$. Hence, $D = 1$ and this is axiom 3. This proves both the existence of an n-rowed determinant as well as the truth of formula (13), the so-called development of a determinant according to its i-th row. (13) may be generalized as follows: In our determinant replace the i-th row by the j-th row and develop according to this new row. For $i \neq j$ that determinant is 0 and for $i = j$ it is D:

$$(14)\quad a_{j1}A_{i1} + a_{j2}A_{i2} + \ldots + a_{jn}A_{in} = \begin{cases} D \text{ for } j = i \\ 0 \text{ for } j \neq i \end{cases}$$

If we interchange the rows and the columns we get the following formula:

$$(15)\quad a_{1h}A_{1k} + a_{2h}A_{2k} + \ldots + a_{nh}A_{nk} = \begin{cases} D \text{ for } h = k \\ 0 \text{ for } h \neq k \end{cases}$$

Now let A represent an n-rowed and B an m-rowed square matrix. By $|A|$, $|B|$ we mean their determinants. Let C be a matrix of n rows and m columns and form the square matrix of $n + m$ rows

$$(16)\quad \begin{pmatrix} A & C \\ 0 & B \end{pmatrix}$$

where 0 stands for a zero matrix with m rows and n columns. If we consider the determinant of the matrix (16) as a function of the columns of A only, it satisfies obviously the first two of our axioms. Because of (12) its value is $c \cdot |A|$ where c is the determinant of (16) after substituting unit vectors for the columns of A. This c still depends on B and considered as function of the rows of B satisfies the first two axioms. Therefore the determinant of (16) is $d \cdot |A| \cdot |B|$ where d is the special case of the determinant of (16) with unit vectors for the columns of A as well as of B. Subtracting multiples of the columns of A from C we can replace C by 0. This shows $d = 1$ and hence the formula

$$(17) \qquad \begin{vmatrix} A & C \\ 0 & B \end{vmatrix} = |A| \cdot |B|.$$

In a similar fashion we could have shown

$$(18) \qquad \begin{vmatrix} A & 0 \\ C & B \end{vmatrix} = |A| \cdot |B|.$$

The formulas (17), (18) are special cases of a general theorem by Lagrange that can be derived from them. We refer the reader to any textbook on determinants since in most applications (17) and (18) are sufficient.

We now investigate what it means for a matrix if its determinant is zero. We can easily establish the following facts:

a) If $A_1, A_2, \ldots, A_n$ are linearly dependent, then $D(A_1, A_2, \ldots, A_n) = 0$. Indeed one of the vectors, say A_k, is then a linear combination of the other columns; subtracting this linear combination from the column A_k reduces it to 0 and so $D = 0$.

b) If any vector B can be expressed as linear combination of $A_1, A_2, \ldots, A_n$ then $D(A_1, A_2, \ldots, A_n) \neq 0$. Returning to (6) and (10) we may select the values for b_{ik} in such a fashion that every $A_i' = U_i$. For this choice the left side in (10) is 1 and hence $D(A_1, A_2, \ldots, A_n)$ on the right side $\neq 0$.

c) Let $A_1, A_2, \ldots, A_n$ be linearly independent and B any other vector. If we go back to the components in the equation $A_1x_1 + A_2x_2 + \ldots + A_nx_n + By = 0$ we obtain n linear homogeneous equations in the $n + 1$ unknowns $x_1, x_2, \ldots, x_n$, y. Consequently, there is a non-trivial solution. y must be $\neq 0$ or else the $A_1, A_2, \ldots, A_n$ would be linearly dependent. But then we can compute B out of this equation as a linear combination of $A_1, A_2, \ldots, A_n$.

Combining these results we obtain:

A determinant vanishes if and only if the column vectors (or the row vectors) are linearly dependent.

Another way of expressing this result is:

The set of n linear homogeneous equations

$$a_{i1}x_1 + a_{i2}x_2 + \dots + a_{in}x_n = 0 \qquad (i = 1, 2, \dots, n)$$

in n unknowns has a non-trivial solution if and only if the determinant of the coefficients is zero.

Another result that can be deduced is:

If $A_1, A_2, \dots, A_n$ are given, then their linear combinations can represent any other vector B if and only if $D(A_1, A_2, \dots, A_n) \neq 0$.

Or:

The set of linear equations

$$(19) \quad a_{i1}x_1 + a_{i2}x_2 + \dots + a_{in}x_n = b_i \qquad (i = 1, 2, \dots, n)$$

has a solution for arbitrary values of the b_i if and only if the determinant of a_{ik} is $\neq 0$. In that case the solution is unique.

We finally express the solution of (19) by means of determinants if the determinant D of the a_{ik} is $\neq 0$.

We multiply for a given k the i-th equation with A_{ik} and add the equations. (15) gives

$$(20) \quad D \cdot x_k = A_{1k}b_1 + A_{2k}b_2 + \dots + A_{nk}b_n \qquad (k = 1, 2, \dots, n)$$

and this gives x_k. The right side in (12) may also be written as the determinant obtained from D by replacing the k-th column by $b_1, b_2, \dots, b_n$. The rule thus obtained is known as Cramer's rule.

II FIELD THEORY

A. Extension Fields.

If E is a field and F a subset of E which, under the operations of addition and multiplication in E, itself forms a field, that is, if F is a subfield of E, then we shall call E an extension of F. The relation of being an extension of F will be briefly designated by $F \subset E$. If $\alpha, \beta, \gamma, \ldots$ are elements of E, then by $F(\alpha, \beta, \gamma, \ldots)$ we shall mean the set of elements in E which can be expressed as quotients of polynomials in $\alpha, \beta, \gamma, \ldots$ with coefficients in F. It is clear that $F(\alpha, \beta, \gamma, \ldots)$ is a field and is the smallest extension of F which contains the elements $\alpha, \beta, \gamma, \ldots$. We shall call $F(\alpha, \beta, \gamma, \ldots)$ the field obtained after the adjunction of the elements $\alpha, \beta, \gamma, \ldots$ to F, or the field generated out of F by the elements $\alpha, \beta, \gamma, \ldots$. In the sequel all fields will be assumed commutative.

If $F \subset E$, then ignoring the operation of multiplication defined between the elements of E, we may consider E as a vector space over F. By the degree of E over F, written (E/F), we shall mean the dimension of the vector space E over F. If (E/F) is finite, E will be called a finite extension.

THEOREM 6. If F, B, E are three fields such that $F \subset B \subset E$, then

$$(E/F) = (B/F)(E/B).$$

Let $A_1, A_2, \ldots, A_r$ be elements of E which are linearly independent with respect to B and let $C_1, C_2, \ldots, C_s$ be elements of B which are independent with respect to F. Then the products $C_i A_j$ where $i = 1, 2, \ldots, s$ and $j = 1, 2, \ldots, r$ are elements of E which are independent with respect to F. For if $\sum_{i,j} a_{ij} C_i A_j = 0$, then $\sum_j (\sum_i a_{ij} C_i) A_j$ is a linear combination of the A_j with coefficients in B and because the A_j were independent with respect to B we have $\sum_i a_{ij} C_i = 0$ for each j. The independence of the C_i with respect to F then requires that each $a_{ij} = 0$. Since there are $r \cdot s$ elements $C_i A_j$ we have shown that for each $r \leq (E/B)$ and $s \leq (B/F)$ the degree $(E/F) \geq r \cdot s$. Therefore, $(E/F) \geq (B/F)(E/B)$. If one of the latter numbers is infinite, the theorem follows. If both (E/B) and (B/F) are finite, say r and s respectively, we may suppose that the A_j and the C_i are generating systems of E and B respectively, and we show that the set of products $C_i A_j$ is a generating system of E over F. Each $A \in E$ can be expressed linearly in terms of the A_j with coefficients in B. Thus, $A = \Sigma B_j A_j$. Moreover, each B_j being an element of B can be expressed linearly with coefficients in F in terms of the C_i, i.e., $B_j = \Sigma a_{ij} C_i$, $j = 1, 2, \ldots, r$. Thus, $A = \Sigma a_{ij} C_i A_j$ and the $C_i A_j$ form an independent generating system of E over F.

Corollary. If $F \subset F_1 \subset F_2 \subset \ldots \subset F_n$, then $(F_n/F) = (F_1/F) \cdot (F_2/F_1) \ldots (F_n/F_{n-1})$.

B. Polynomials.

An expression of the form $a_0 x^n + a_1 x^{n-1} + \ldots + a_n$ is called a polynomial in F of degree n if the coefficients

$a_o, \ldots, a_n$ are elements of the field F and $a_o \neq 0$. Multiplication and addition of polynomials are performed in the usual way [1].

A polynomial in F is called reducible in F if it is equal to the product of two polynomials in F each of degree at least one. Polynomials which are not reducible in F are called irreducible in F.

If $f(x) = g(x) \cdot h(x)$ is a relation which holds between the polynomials $f(x)$, $g(x)$, $h(x)$ in a field F, then we shall say that $g(x)$ divides $f(x)$ in F, or that $g(x)$ is a factor of $f(x)$. It is readily seen that the degree of $f(x)$ is equal to the sum of the degrees of $g(x)$ and $h(x)$, so that if neither $g(x)$ nor $h(x)$ is a constant then each has a degree less than $f(x)$. It follows from this that by a finite number of factorizations a polynomial can always be expressed as a product of irreducible polynomials in a field F.

For any two polynomials $f(x)$ and $g(x)$ the division algorithm holds, i.e., $f(x) = q(x) \cdot g(x) + r(x)$ where $q(x)$ and $r(x)$ are unique polynomials in F and the degree of $r(x)$ is less than that of $g(x)$. This may be shown by the same argument as the reader met in elementary algebra in the case of the field of real or complex numbers. We also see that $r(x)$ is the uniquely determined polynomial of a degree less than that of $g(x)$ such that $f(x) - r(x)$ is divisible by $g(x)$. We shall call $r(x)$ the remainder of $f(x)$.

---

1) If we speak of the set of all polynomials of degree lower than n, we shall agree to include the polynomial 0 in this set, though it has no degree in the proper sense.

Also, in the usual way, it may be shown that if α is a root of the polynomial f(x) in F than $x - \alpha$ is a factor of f(x), and as a consequence of this that a polynomial in a field cannot have more roots in the field than its degree.

Lemma. If f(x) is an irreducible polynomial of degree n in F, then there do not exist two polynomials each of degree less than n in F whose product is divisible by f(x).

Let us suppose to the contrary that g(x) and h(x) are polynomials of degree less than n whose product is divisible by f(x). Among all polynomials occurring in such pairs we may suppose g(x) has the smallest degree. Then since f(x) is a factor of $g(x) \cdot h(x)$ there is a polynomial k(x) such that

$$k(x) \cdot f(x) = g(x) \cdot h(x)$$

By the division algorithm,

$$f(x) = q(x) \cdot g(x) + r(x)$$

where the degree of r(x) is less than that of g(x) and $r(x) \neq 0$ since f(x) was assumed irreducible. Multiplying

$$f(x) = q(x) \cdot g(x) + r(x)$$

by h(x) and transposing, we have

$$r(x) \cdot h(x) = f(x) \cdot h(x) - q(x) \cdot g(x) \cdot h(x) = f(x) \cdot h(x) - q(x) \cdot k(x) \cdot f(x)$$

from which it follows that $r(x) \cdot h(x)$ is divisible by f(x). Since r(x) has a smaller degree than g(x), this last is in contradiction to the choice of g(x), from which the lemma follows.

As we saw, many of the theorems of elementary algebra hold in any field F. However, the so-called Fundamental Theorem of Algebra, at least in its customary form, does not hold. It will be replaced by a theorem due to Kronecker which guarantees for a given polynomial in F the existence of an extension field in which the polynomial has a root. We shall also show that, in a given field, a polynomial can not only be factored into irreducible factors, but that this factorization is unique up to a constant factor. The uniqueness depends on the theorem of Kronecker.

C. Algebraic Elements.

Let F be a field and E an extension field of F. If α is an element of E we may ask whether there are polynomials with coefficients in F which have α as root. α is called algebraic with respect to F if there are such polynomials. Now let α be algebraic and select among all polynomials in F which have α as root one, $f(x)$, of lowest degree.

We may assume that the highest coefficient of $f(x)$ is 1. We contend that this $f(x)$ is uniquely determined, that it is irreducible and that each polynomial in F with the root α is divisible by $f(x)$. If, indeed, $g(x)$ is a polynomial in F with $g(\alpha) = 0$, we may divide $g(x) = f(x)q(x) + r(x)$ where $r(x)$ has a degree smaller than that of $f(x)$. Substituting $x = \alpha$ we get $r(\alpha) = 0$. Now $r(x)$ has to be identically 0 since otherwise $r(x)$ would have the root α and be of lower degree than $f(x)$. So $g(x)$ is divisible by $f(x)$. This also shows the uniqueness of $f(x)$. If $f(x)$ were not irreducible, one of the factors would have to vanish for $x = \alpha$ contradicting again the choice of $f(x)$.

We consider now the subset E_o of the following elements θ of E:

$$\theta = g(\alpha) = c_o + c_1\alpha + c_2\alpha^2 + \dots + c_{n-1}\alpha^{n-1}$$

where $g(x)$ is a polynomial in F of degree less than n (n being the degree of $f(x)$). This set E_o is closed under addition and multiplication. The latter may be verified as follows:

If $g(x)$ and $h(x)$ are two polynomials of degree less than n we put $g(x)h(x) = q(x)f(x) + r(x)$ and hence $g(\alpha)h(\alpha) = r(\alpha)$. Finally we see that the constants $c_o, c_1, \dots, c_{n-1}$ are uniquely determined by the element θ. Indeed two expressions for the same θ would lead after subtracting to an equation for α of lower degree than n.

We remark that the internal structure of the set E_o does not depend on the nature of α but only on the irreducible $f(x)$. The knowledge of this polynomial enables us to perform the operations of addition and multiplication in our set E_o. We shall see very soon that E_o is a field; in fact, E_o is nothing but the field $F(\alpha)$. As soon as this is shown we have at once the degree, $(F(\alpha)/F)$, determined as n, since the space $F(\alpha)$ is generated by the linearly independent $1, \alpha, \alpha^2, \dots, \alpha^{n-1}$.

We shall now try to imitate the set E_o without having an extension field E and an element α at our disposal. We shall assume only an irreducible polynomial

$$f(x) = x^n + a_{n-1}x^{n-1} + \dots + a_o$$

as given.

We select a symbol ξ and let E_1 be the set of all formal polynomials

$$g(\xi) = c_o + c_1\xi + \dots + c_{n-1}\xi^{n-1}$$

of a degree lower than n. This set forms a group under addition. We now introduce besides the ordinary multiplication a new kind of multiplication of two elements $g(\xi)$ and $h(\xi)$ of E_1 denoted by $g(\xi) \times h(\xi)$. It is defined as the remainder $r(\xi)$ of the ordinary product $g(\xi)h(\xi)$ under division by $f(\xi)$. We first remark that any product of m terms $g_1(\xi), g_2(\xi), \ldots, g_m(\xi)$ is again the remainder of the ordinary product $g_1(\xi)g_2(\xi)\ldots g_m(\xi)$. This is true by definition for $m = 2$ and follows for every m by induction if we just prove the easy lemma: The remainder of the product of two remainders (of two polynomials) is the remainder of the product of these two polynomials. This fact shows that our new product is associative and commutative and also that the new product $g_1(\xi) \times g_2(\xi) \times \ldots \times g_m(\xi)$ will coincide with the old product $g_1(\xi)g_2(\xi)\ldots g_m(\xi)$ if the latter does not exceed n in degree. The distributive law for our multiplication is readily verified.

The set E_1 contains our field F and our multiplication in E_1 has for F the meaning of the old multiplication. One of the polynomials of E_1 is ξ. Multiplying it i-times with itself, clearly will just lead to ξ^i as long as $i < n$. For $i = n$ this is not any more the case since it leads to the remainder of the polynomial ξ^n.
This remainder is

$$\xi^n - f(\xi) = -a_{n-1}\xi^{n-1} - a_{n-2}\xi^{n-2} - \ldots - a_o.$$

We now give up our old multiplication altogether and keep only the new one; we also change notation, using the point (or juxtaposition) as symbol for the new multiplication.

Computing in this sense

$$c_o + c_1\xi + c_2\xi^2 + \ldots + c_{n-1}\xi^{n-1}$$

will readily lead to this element, since all the degrees

involved are below n. But

$$\xi^n = -a_{n-1}\xi^{n-1} - a_{n-2}\xi^{n-2} - \dots - a_o.$$

Transposing we see that $f(\xi) = 0$.

We thus have constructed a set E_1 and an addition and multiplication in E_1 that already satisfies most of the field axioms. E_1 contains F as subfield and ξ satisfies the equation $f(\xi) = 0$. We next have to show: If $g(\xi) \neq 0$ and $h(\xi)$ are given elements of E_1, there is an element

$$X(\xi) = x_o + x_1\xi + \dots + x_{n-1}\xi^{n-1}$$

in E_1 such that

$$g(\xi) \cdot X(\xi) = h(\xi).$$

To prove it we consider the coefficients x_i of $X(\xi)$ as unknowns and compute nevertheless the product on the left side, always reducing higher powers of ξ to lower ones. The result is an expression $L_o + L_1\xi + \dots + L_{n-1}\xi^{n-1}$ where each L_i is a linear combination of of the x_i with coefficients in F. This expression is to be equal to $h(\xi)$; this leads to the n equations with n unknowns:

$$L_o = b_o, \; L_1 = b_1, \; \dots, \; L_{n-1} = b_{n-1}$$

where the b_i are the coefficients of $h(\xi)$. This system will be soluble if the corresponding homogeneous equations

$$L_o = 0, \; L_1 = 0, \; \dots, \; L_{n-1} = 0$$

have only the trivial solution.

The homogeneous problem would occur if we should ask for the set of elements $X(\xi)$ satisfying $g(\xi) \cdot X(\xi) = 0$. Going back for a moment to the old multiplication this would mean that the ordinary product $g(\xi)\, X(\xi)$ has the remainder 0, and is therefore divisible by $f(\xi)$. According to the lemma, page 24, this is only possible for $X(\xi) = 0$.

Therefore E_1 is a field.

Assume now that we have also our old extension E with a root α of $f(x)$, leading to the set E_0. We see that E_0 has in a certain sense the same structure as E_1 if we map the element $g(\xi)$ of E_1 onto the element $g(\alpha)$ of E_0. This mapping will have the property that the image of a sum of elements is the sum of the images, and the image of a product is the product of the images.

Let us therefore define: A mapping σ of one field onto another which is one to one in both directions such that $\sigma(\alpha+\beta) = \sigma(\alpha) + \sigma(\beta)$ and $\sigma(\alpha\cdot\beta) = \sigma(\alpha)\cdot\sigma(\beta)$ is called an isomorphism. If the fields in question are not distinct – i.e., are both the same field – the isomorphism is called an automorphism. Two fields for which there exists an isomorphism mapping one on another are called isomorphic. If not every element of the image field is the image under σ of an element in the first field, then σ is called an isomorphism of the first field into the second. Under each isomorphism it is clear that $\sigma(0) = 0$ and $\sigma(1) = 1$.

We see that E_0 is also a field and that it is isomorphic to E_1.

We now mention a few theorems that follow from our discussion:

THEOREM 7. (Kronecker). If $f(x)$ is a polynomial in a field F, there exists an extension E of F in which $f(x)$ has a root.

Proof: Construct an extension field in which an irreducible factor of $f(x)$ has a root.

THEOREM 8. Let σ be an isomorphism mapping a field F on a field F'. Let $f(x)$ be an irreducible polynomial in F and $f'(x)$ the corresponding polynomial in F'. If $E = F(\beta)$ and $E' = F'(\beta')$ are extensions of F and F', respectively, where $f(\beta) = 0$ in E and $f'(\beta') = 0$ in E' then σ can be extended to an isomorphism between E and E'.

Proof: E and E' are both isomorphic to E_0.

D. Splitting Fields.

If F, B and E are three fields such that $F \subset B \subset E$, then we shall refer to B as an intermediate field.

If E is an extension of a field F in which a polynomial $p(x)$ in F can be factored into linear factors, and if $p(x)$ can not be so factored in any intermediate field, then we call E a splitting field for $p(x)$. Thus, if E is a splitting field of $p(x)$, the roots of $p(x)$ generate E.

A splitting field is of finite degree since it is constructed by a finite number of adjunctions of algebraic elements, each defining an extension field of finite degree. Because of the corollary on page 22, the total degree is finite.

THEOREM 9. If $p(x)$ is a polynomial in a field F, there exists a splitting field E of $p(x)$.

We factor $p(x)$ in F into irreducible factors $f_1(x) \ldots f_r(x) = p(x)$. If each of these is of the first degree then F itself is the required splitting field. Suppose then that $f_1(x)$ is of degree higher than the first. By

Theorem 7 there is an extension F_1 of F in which $f_1(x)$ has a root. Factor each of the factors $f_1(x), \ldots, f_r(x)$ into irreducible factors in F_1 and proceed as before. We finally arrive at a field in which p(x) can be split into linear factors. The field generated out of F by the roots of p(x) is the required splitting field.

The following theorem asserts that up to isomorphisms, the splitting field of a polynomial is unique.

THEOREM 10. Let σ be an isomorphism mapping the field F on the field F', Let p(x) be a polynomial in F and p'(x) the polynomial in F' with coefficients corresponding to those of p(x) under σ. Finally, let E be a splitting field of p(x) and E' a splitting field of p'(x). Under these conditions the isomorphism σ can be extended to an isomorphism between E and E'.

If f(x) is an irreducible factor of p(x) in F, then E contains a root of f(x). For let $p(x)=(x-\alpha_1)(x-\alpha_2)\ldots(x-\alpha_s)$ be the splitting of p(x) in E. Then $(x-\alpha_1)(x-\alpha_2)\ldots(x-\alpha_s) = f(x)\ g(x)$. We consider f(x) as a polynomial in E and construct the extension field $B = E(\alpha)$ in which $f(\alpha) = 0$. Then $(\alpha-\alpha_1)\cdot(\alpha-\alpha_2)\cdot\ldots\cdot(\alpha-\alpha_s) = f(\alpha)\cdot g(\alpha) = 0$ and $\alpha-\alpha_i$ being elements of the field B can have a product equal to 0 only if for one of the factors, say the first, we have $\alpha-\alpha_1 = 0$. Thus, $\alpha = \alpha_1$, and α_1 is a root of f(x).

Now in case all roots of p(x) are in F, then E = F and p(x) can be split in F. This factored form has an image in F' which is a splitting of p'(x), since the isomorphism σ preserves all operations of addition and multiplication in the process of multiplying out the factors of $p(x)$ and collecting to get the original form. Since $p'(x)$ can be split in F', we must have $F' = E'$. In this case, σ itself is the required extension and the theorem is proved if all roots of $p(x)$ are in F.

We proceed by complete induction. Let us suppose the theorem proved for all cases in which the number of roots of $p(x)$ outside of F is less than $n > 1$, and suppose that $p(x)$ is a polynomial having n roots outside of F. We factor $p(x)$ into irreducible factors in F; $p(x) = f_1(x)\, f_2(x) \ldots f_m(x)$. Not all of these factors can be of degree 1, since otherwise $p(x)$ would split in F, contrary to assumption. Hence, we may suppose the degree of $f_1(x)$ to be $r > 1$. Let $f'_1(x) \cdot f'_2(x) \ldots f'_m(x) = p'(x)$ be the factorization of $p'(x)$ into the polynomials corrrespondng to $f_1(x), \ldots, f_m(x)$ under σ. $f'_1(x)$ is irreducible in F', for a factorization of $f'_1(x)$ in F' would induce[1] under σ^{-1} a factorization of $f_1(x)$, which was however taken to be irreducible.

By Theorem 8, the isomorphism σ can be extended to an isomorphism σ_1, between the fields $F(\alpha)$ and $F'(\alpha')$.

Since $F \subset F(\alpha)$, $p(x)$ is a polynomial in $F(\alpha)$ and E is a splitting field for $p(x)$ in $F(\alpha)$. Similarly for $p'(x)$. There are now less than n roots of $p(x)$ outside the new ground field $F(\alpha)$. Hence by our inductive assumption σ_1 can be extended from an isomorphism between $F(\alpha)$ and $F'(\alpha')$ to an isomorphism σ_2 between E and E'. Since σ_1 is an extension of σ, and σ_2 an extension of σ_1, we conclude σ_2 is an extension of σ and the theorem follows.

---

1) See page 38 for the definition of σ^{-1}.

Corollary. If $p(x)$ is a polynomial in a field F, then any two splitting fields for $p(x)$ are isomorphic.

This follows from Theorem 10 if we take $F = F'$ and σ to be the identity mapping, i.e., $\sigma(x) = x$.

As a consequence of this corollary we see that we are justified in using the expression "the splitting field of $p(x)$" since any two differ only by an isomorphism. Thus, if $p(x)$ has repeated roots in one splitting field, so also in any other splitting field it will have repeated roots. The statement "$p(x)$ has repeated roots" will be significant without reference to a particular splitting field.

E. Unique Decomposition of Polynomials into Irreducible Factors.

THEOREM 11. If $p(x)$ is a polynomial in a field F, and if $p(x) = p_1(x)\cdot p_2(x)\cdot \ldots \cdot p_r(x) = q_1(x)\cdot q_2(x)\cdot \ldots \cdot q_s(x)$ are two factorizations of $p(x)$ into irreducible polynomials each of degree at least one, then $r = s$ and after a suitable change in the order in which the q's are written, $p_i(x) = c_i q_i(x)$, $i = 1, 2, \ldots, r$, and $c_i \in F$.

Let $F(\alpha)$ be an extension of F in which $p_1(\alpha) = 0$. We may suppose the leading coefficients of the $p_i(x)$ and the $q_i(x)$ to be 1, for, by factoring out all leading coefficients and combining, the constant multiplier on each side of the equation must be the leading coefficient of $p(x)$ and hence can be divided out of both sides of the equation. Since $0 = p_1(\alpha)\cdot p_2(\alpha)\cdot \ldots \cdot p_r(\alpha) = p(\alpha) = q_1(\alpha)\cdot \ldots \cdot q_s(\alpha)$ and since a product of elements of $F(\alpha)$ can be 0 only if one of these is 0, it follows that one of the $q_i(\alpha)$, say $q_1(\alpha)$, is 0. This gives (see page 25) $p_1(x) = q_1(x)$. Thus $p_1(x) . p_2(x) . \ldots . p_r(x)$
$= p_1(x)\cdot q_2(x)\cdot \ldots \cdot q_s(x)$ or

$p_1(x) \cdot [p_2(x) \cdot \ldots \cdot p_r(x) - q_2(x) \cdot \ldots \cdot q_s(x)] = 0$. Since the product of two polynomials is 0 only if one of the two is the 0 polynomial, it follows that the polynomial within the brackets is 0 so that $p_2(x) \cdot \ldots \cdot p_r(x) = q_2(x) \cdot \ldots \cdot q_s(x)$. If we repeat the above argument r times we obtain $p_i(x) = q_i(x)$, $i = 1, 2, \ldots, r$. Since the remaining q's must have a product 1, it follows that $r = s$.

F. Group Characters.

If G is a multiplicative group, F a field and σ a homomorphism mapping G into F, then σ is called a character of G in F. By homomorphism is meant a mapping σ such that for α, β any two elements of G, $\sigma(a) \cdot \sigma(\beta) = \sigma(\alpha \cdot \beta)$ and $\sigma(\alpha) \neq 0$ for any α.
(If $\sigma(\alpha) = 0$ for one element α, then $\sigma(x) = 0$ for each $x \in G$, since $\sigma(\alpha y) = \sigma(\alpha) \cdot \sigma(y) = 0$ and αy takes all values in G when y assumes all values in G).

The characters $\sigma_1, \sigma_2, \ldots, \sigma_n$ are called dependent if there exist elements $a_1, a_2, \ldots, a_n$ not all zero in F such that $a_1\sigma_1(x) + a_2\sigma_2(x) + \ldots + a_n\sigma_n(x) = 0$ for each $x \in G$. Such a dependence relation is called non-trivial. If the characters are not dependent they are called independent.

THEOREM 12. If G is a group and $\sigma_1, \sigma_2, \ldots, \sigma_n$ are n mutually distinct characters of G in a field F, then $\sigma_1, \sigma_2, \ldots, \sigma_n$ are independent.

One character cannot be dependent, since $a_1\sigma_1(x) = 0$ implies $a_1 = 0$ due to the assumption that $\sigma_1(x) \neq 0$. Suppose $n > 1$.

We make the inductive assumption that no set of less than n distinct characters is dependent. Suppose now that

$a_1\sigma_1(x) + a_2\sigma_2(x) + \ldots + a_n\sigma_n(x) = 0$ is a non-trivial dependence between the σ's. None of the elements a_i is zero, else we should have a dependence between less than n characters contrary to our inductive assumption. Since σ_1 and σ_n are distinct, there exists an element α in G such that $\sigma_1(\alpha) \neq \sigma_n(\alpha)$. Multiply the relation between the σ's by a_n^{-1} . We obtain a relation

$$(*) \quad b_1\sigma_1(x) + \ldots + b_{n-1}\,\sigma_{n-1}(x) + \sigma_n(x) = 0,\ b_i = a_n^{-1}\,a_i \neq 0.$$

Replace in this relation x by αx. We have

$$b_1\sigma_1(\alpha)\sigma_1(x) + \ldots + b_{n-1}\,\sigma_{n-1}(\alpha)\sigma_{n-1}(x) + \sigma_n(\alpha)\sigma_n(x) = 0,$$

or $\sigma_n(\alpha)^{-1}\,b_1\sigma_1(\alpha)\sigma_1(x) + \ldots + \sigma_n(x) = 0.$

Subtracting the latter from (*) we have

$$(**) \quad [b_1 - \sigma_n(\alpha)^{-1}b_1\sigma_1(\alpha)]\sigma_1(x) + \ldots + c_{n-1}\sigma_{n-1}(x) = 0.$$

The coefficient of $\sigma_1(x)$ in this relation is not 0, otherwise we should have $b_1 = \sigma_n(\alpha)^{-1}b_1\sigma_1(a)$, so that

$$\sigma_n(\alpha)b_1 = b_1\sigma_1(\alpha) = \sigma_1(\alpha)b_1$$

and since $b_1 \neq 0$, we get $\sigma_n(\alpha) = \sigma_1(\alpha)$ contrary to the choice of α. Thus, (**) is a non-trivial dependence between $\sigma_1, \sigma_2, \ldots, \sigma_{n-1}$ which is contrary to our inductive assumption.

Corollary. If E and E' are two fields, and $\sigma_1, \sigma_2, \ldots, \sigma_n$ are n mutually distinct isomorphisms mapping E into E', then $\sigma_1, \ldots, \sigma_n$ are independent. (Where "independent" again means there exists no non-trivial dependence $a_1\sigma_1(x) + \ldots + a_n\sigma_n(x) = 0$ which holds for every $x \in E$).

This follows from Theorem 12, since E without the 0 is a group and the σ's defined in this group are mutually distinct characters.

If $\sigma_1, \sigma_2, \ldots, \sigma_n$ are isomorphisms of a field E into a field E', then each element a of E such that $\sigma_1(a) = \sigma_2(a) = \ldots = \sigma_n(a)$ is called a fixed point of E under $\sigma_1, \sigma_2, \ldots, \sigma_n$. This name is chosen because in the case where the σ's are automorphisms and σ_1 is the identity, i.e., $\sigma_1(x) = x$, we have $\sigma_i(x) = x$ for a fixed point.

Lemma. The set of fixed points of E is a subfield of E. We shall call this subfield the fixed field.

For if a and b are fixed points, then

$\sigma_i(a + b) = \sigma_i(a) + \sigma_i(b) = \sigma_j(a) + \sigma_j(b) = \sigma_j(a + b)$ and

$\sigma_i(a \cdot b) = \sigma_i(a) \cdot \sigma_i(b) = \sigma_j(a) \cdot \sigma_j(b) = \sigma_j(a \cdot b)$.

Finally from $\sigma_i(a) = \sigma_j(a)$ we have $(\sigma_j(a))^{-1} = (\sigma_i(a))^{-1}$

$= \sigma_i(a^{-1}) = \sigma_j(a^{-1})$.

Thus, the sum and product of two fixed points is a fixed point, and the inverse of a fixed point is a fixed point. Clearly, the negative of a fixed point is a fixed point.

THEOREM 13. If $\sigma_1, \ldots, \sigma_n$ are n mutually distinct isomorphisms of a field E into a field E', and if F is the fixed field of E, then $(E/F) \geq n$.

Suppose to the contrary that $(E/F) = r < n$. We shall show that we are led to a contradiction. Let $\omega_1, \omega_2, \ldots, \omega_r$ be a generating system of E over F. In the homogeneous linear equations

$$\sigma_1(\omega_1)x_1 + \sigma_2(\omega_1)x_2 + \dots + \sigma_n(\omega_1)x_n = 0$$
$$\sigma_1(\omega_2)x_1 + \sigma_2(\omega_2)x_2 + \dots + \sigma_n(\omega_2)x_n = 0$$
$$\dots\dots\dots\dots\dots\dots\dots\dots\dots$$
$$\sigma_1(\omega_r)x_1 + \sigma_2(\omega_r)x_2 + \dots + \sigma_n(\omega_r)x_n = 0$$

there are more unknowns than equations so that there exists a non-trivial solution which, we may suppose, $x_1, x_2, \dots, x_n$ denotes. For any element α in E we can find $a_1, a_2, \dots, a_r$ in F such that $\alpha = a_1\omega_1 + \dots + a_r\omega_r$. We multiply the first equation by $\sigma_1(a_1)$, the second by $\sigma_1(a_2)$, and so on. Using that $a_i \in F$, hence that $\sigma_1(a_i) = \sigma_j(a_i)$ and also that $\sigma_j(a_i)\ \sigma_j(\omega_i) = \sigma_j(a_i\omega_i)$, we obtain

$$\sigma_1(a_1\omega_1)x_1 + \dots + \sigma_n(a_1\omega_1)x_n = 0$$
$$\dots\dots\dots\dots\dots\dots\dots\dots$$
$$\sigma_1(a_r\omega_r)x_1 + \dots + \sigma_n(a_r\omega_r)x_n = 0.$$

Adding these last equations and using

$$\sigma_i(a_1\omega_1) + \sigma_i(a_2\omega_2) + \dots + \sigma_i(a_r\omega_r) = \sigma_i(a_1\omega_1 + \dots + a_r\omega_r) = \sigma_i(\alpha)$$

we obtain

$$\sigma_1(\alpha)x_1 + \sigma_2(\alpha)x_2 + \dots + \sigma_n(\alpha)x_n = 0.$$

This, however, is a non-trivial dependence relation between $\sigma_1, \sigma_2, \dots, \sigma_n$ which cannot exist according to the corollary of Theorem 12.

Corollary. If $\sigma_1, \sigma_2, \dots, \sigma_n$ are automorphisms of the field E, and F is the fixed field, then $(E/F) \geq n$.

If F is a subfield of the field E, and σ an automorphism of E, we shall say that σ leaves F fixed if for each element a of F, $\sigma(a) = a$.

If σ and τ are two automorphisms of E, then the mapping $\sigma(\tau(x))$ written briefly $\sigma\tau$ is an automorphism, as the reader may readily verify. [E.g., $\sigma\tau(x \cdot y) = \sigma(\tau(x \cdot y)) = \sigma(\tau(x) \cdot \tau(y)) = \sigma(\tau(x)) \cdot \sigma(\tau(y))$]. We shall call $\sigma\tau$ the product of σ and τ. If σ is an automorphism ($\sigma(x) = y$), then we shall call σ^{-1} the mapping of y into x, i.e., $\sigma^{-1}(y) = x$ the inverse of σ. The reader may readily verify that σ^{-1} is an automorphism. The automorphism $I(x) = x$ shall be called the unit automorphism.

Lemma. If E is an extension field of F, the set G of automorphisms which leave F fixed is a group.

The product of two automorphisms which leave F fixed clearly leaves F fixed. Also, the inverse of any automorphism in G is in G.

The reader will observe that G, the set of automorphisms which leave F fixed, does not necessarily have F as its fixed field. It may be that certain elements in E which do not belong to F are left fixed by every automorphism which leaves F fixed. Thus, the fixed field of G may be larger than F.

G. Applications and Examples to Theorem 13.

Theorem 13 is very powerful as the following examples show:

1) Let k be a field and consider the field $E = k(x)$ of all rational functions of the variable x. If we map each of the functions $f(x)$ of E onto $f(\frac{1}{x})$ we obviously obtain an automorphism of E. Let us consider the following six automorphisms where $f(x)$ is mapped onto $f(x)$ (identity), $f(1-x)$, $f(\frac{1}{x})$, $f(1-\frac{1}{x})$, $f(\frac{1}{1-x})$ and $f(\frac{x}{x-1})$ and call F the

fixed point field. F consists of all rational functions satisfying

(1) $$f(x) = f(1-x) = f(\frac{1}{x}) = f(1-\frac{1}{x}) = f(\frac{1}{1-x}) = f(\frac{x}{x-1}).$$

It suffices to check the first two equalities, the others being consequences. The function

(2) $$I = I(x) = \frac{(x^2 - x+1)^3}{x^2(x-1)^2}$$

belongs to F as is readily seen. Hence, the field $S = k(I)$ of all rational functions of I will belong to F.

We contend: $F = S$ and $(E/F) = 6$.

Indeed, from Theorem 13 we obtain $(E/F) \geq 6$. Since $S \subset F$ it suffices to prove $(E/S) \leq 6$. Now $E = S(x)$. It is thus sufficient to find some 6-th degree equation with coefficients in S satisfied by x. The following one is obviously satisfied;

$$(x^2 - x+1)^3 - I \cdot x^2(x-1)^2 = 0.$$

The reader will find the study of these fields a profitable exercise. At a later occasion he will be able to derive all intermediate fields.

2) Let k be a field and $E = k(x_1, x_2, \ldots, x_n)$ the field of all rational functions of n variables $x_1, x_2, \ldots, x_n$. If $(\nu_1, \nu_2, \ldots, \nu_n)$ is a permutation of $(1, 2, \ldots, n)$ we replace in each function $f(x_1, x_2, \ldots, x_n)$ of E the variable x_1 by x_{ν_1}, x_2 by $x_{\nu_2}, \ldots, x_n$ by x_{ν_n}. The mapping of E onto itself obtained in this way is obviously an automorphism and we may construct $n!$ automorphisms in this fashion (including the identity). Let F be the fixed point field, that is, the set of all so-called "symmetric functions." Theorem 13 shows that $(E/F) \geq n!$. Let us introduce the polynomial:

(3) $$f(t) = (t-x_1)(t-x_2)\ldots(t-x_n) = t^n + a_1 t^{n-1} + \ldots + a_n$$

where $a_1 = -(x_1 + x_2 + \ldots + x_n)$; $a_2 = +(x_1x_2 + x_1x_3 + \ldots + x_{n-1}x$ and more generally a_i is $(-1)^i$ times the sum of all products of i differ- erent variables of the set $x_1, x_2, \ldots, x_n$. The functions $a_1, a_2, \ldots, a_n$ are called the elementary symmetric functions and the field $S = k(a_1, a_2, \ldots, a_n)$ of all rational functions of $a_1, a_2, \ldots, a_n$ is obviously a part of F. Should we suceed in proving $(E/S) \leq n!$ we would have shown $S = F$ and $(E/F) = n!$.

We construct to this effect the following tower of fields:

$$S = S_n \subset S_{n-1} \subset S_{n-2} \ldots \subset S_2 \subset S_1 = E$$

by the definition

(4) $S_n = S$; $S_i = S(x_{i+1}, x_{i+2}, \ldots, x_n) = S_{i+1}(x_{i+1})$.

It would be sufficient to prove $(S_{i-1}/S_i) \leq i$ or that the generator x_i for S_{i-1} out of S_i satisfies an equation of degree i with coefficients in S_i.

Such an equation is easily constructed. Put

(5) $$F_i(t) = \frac{f(t)}{(t-x_{i+1})(t-x_{i+2})\ldots(t-x_n)} = \frac{F_{i+1}(t)}{(t-x_{i+1})}$$

and $F_n(t) = f(t)$. Performing the division we see that $F_i(t)$ is a polynomial in t of degree i whose highest coefficient is 1 and whose coefficients are polynomials in the variables $a_1, a_2, \ldots, a_n$ and $x_{i+1}, x_{i+2}, \ldots, x_n$. Only integers enter as coefficients in these expressions. Now x_i is obviously a root of $F_i(t) = 0$.

Now let $g(x_1, x_2, \ldots, x_n)$ be a <u>polynomial</u> in $x_1, x_2, \ldots, x_n$. Since $F_1(x_1) = 0$ is of first degree in x_1, we can express x_1 as a polynomial of the a_i and of $x_2, x_3, \ldots, x_n$. We introduce this expression in $g(x_1, x_2, \ldots, x_n)$. Since $F_2(x_2) = 0$ we can express x_2^2 or higher powers as polynomials in $x_3, \ldots, x_n$ and the a_i. Since $F_3(x_3) = 0$ we can express x_3^3 and higher powers as polynomials of $x_4, x_5, \ldots, x_n$ and the a_i. Introducing these expressions in $g(x_1, x_2, \ldots, x_n)$ we see that we can express it as a polynomial in the x_ν and the a_ν such that the degree in x_i is below i. So $g(x_1, x_2, \ldots, x_n)$ is a linear combination of the following n ! terms:

$$(6) \qquad x_1^{\nu_1} x_2^{\nu_2} \ldots x_n^{\nu_n} \text{ where each } \nu_i \leq i - 1.$$

The coefficients of these terms are polynomials in the a_i. Since the expressions (6) are linearly independent in S (this is our previous result), the expression is unique.

This is a generalization of the theorem of symmetric functions in its usual form. The latter says that a symmetric polynomial can be written as a polynomial in $a_1, a_2, \ldots, a_n$. Indeed, if $g(x_1, \ldots, x_n)$ is symmetric we have already an expression as linear combination of the terms (6) where only the term 1 corresponding to $\nu_1 = \nu_2 = \ldots = \nu_n = 0$ has a coefficient $\neq 0$ in S, namely, $g(x_1, \ldots, x_n)$. So $g(x_1, x_2, \ldots, x_n)$ is a polynomial in $a_1, a_2, \ldots, a_n$.

But our theorem gives an expression of any polynomial, symmetric or not.

H. Normal Extensions.

An extension field E of a field F is called a normal extension if the group G of automorphisms of E which leave F fixed has F for its fixed field, and (E/F) is finite.

Although the result in Theorem 13 cannot be sharpened in general, there is one case in which the equality sign will always occur, namely, in the case in which $\sigma_1, \sigma_2, \ldots, \sigma_n$ is a set of automorphisms which form a group. We prove

THEOREM 14. If $\sigma_1, \sigma_2, \ldots, \sigma_n$ is a group of automorphisms of a field E and if F is the fixed field of $\sigma_1, \sigma_2, \ldots, \sigma_n$, then $(E/F) = n$.

If $\sigma_1, \sigma_2, \ldots, \sigma_n$ is a group, then the identity occurs, say, $\sigma_1 = I$. The fixed field consists of those elements x which are not moved by any of the σ's, i.e., $\sigma_i(x) = x$, $i = 1, 2, \ldots n$. Suppose that $(E/F) > n$. Then there exist $n + 1$ elements $\alpha_1, \alpha_2, \ldots, \alpha_{n+1}$ of E which are linearly independent with respect to F. By Theorem 1, there exists a non-trivial solution in E to the system of equations

$$(') \quad \begin{aligned} x_1 \sigma_1(\alpha_1) + x_2 \sigma_1(\alpha_2) + \ldots + x_{n+1} \sigma_1(\alpha_{n+1}) &= 0 \\ x_1 \sigma_2(\alpha_1) + x_2 \sigma_2(\alpha_2) + \ldots + x_{n+1} \sigma_2(\alpha_{n+1}) &= 0 \\ \cdots\cdots\cdots\cdots\cdots\cdots\cdots\cdots\cdots\cdots& \\ x_1 \sigma_n(\alpha_1) + x_2 \sigma_n(\alpha_2) + \ldots + x_{n+1} \sigma_n(\alpha_{n+1}) &= 0 \end{aligned}$$

We note that the solution cannot lie in F, otherwise, since σ_1 is the identity, the first equation would be a dependence between $\alpha_1, \ldots, \alpha_{n+1}$.

Among all non-trivial solutions $x_1, x_2, \ldots, x_{n+1}$ we choose one which has the least number of elements different from 0. We may suppose this solution to be $a_1, a_2, \ldots, a_r, 0, \ldots, 0$, where the first r terms are different from 0. Moreover, $r \neq 1$ because $a_1 \sigma_1(\alpha_1) = 0$ implies $a_1 = 0$ since $\sigma_1(\alpha_1) = \alpha_1 \neq 0$. Also, we may suppose $a_r = 1$, since if we multiply the given solution by a_r^{-1} we obtain a new solution in which the r-th term is 1. Thus, we have

(*) $a_1\sigma_i(\alpha_1) + a_2\sigma_i(\alpha_2) + \cdots + a_{r-1}\sigma_i(\alpha_{r-1}) + \sigma_i(\alpha_r) = 0$

for $i = 1, 2, \ldots, n$. Since $a_1, \ldots, a_{r-1}$ cannot all belong to F, one of these, say a_1, is in E but not in F. There is an automorphism σ_k for which $\sigma_k(a_1) \neq a_1$. If we use the fact that $\sigma_1, \sigma_2, \ldots, \sigma_n$ form a group, we see $\sigma_k \cdot \sigma_1, \sigma_k \cdot \sigma_2, \ldots, \sigma_k \cdot \sigma_n$ is a permutation of $\sigma_1, \sigma_2, \ldots, \sigma_n$. Applying σ_k to the expressions in (*) we obtain

$$\sigma_k(a_1)\cdot\sigma_k\sigma_j(\alpha_1) + \cdots + \sigma_k(a_{r-1})\cdot\sigma_k\sigma_j(\alpha_{r-1}) + \sigma_k\sigma_j(\alpha_r) = 0$$

for $j = 1, 2, \ldots, n$, so that from $\sigma_k\sigma_j = \sigma_i$

(**) $\sigma_k(a_1)\sigma_i(\alpha_1) + \cdots + \sigma_k(a_{r-1})\sigma_i(\alpha_{r-1}) + \sigma_i(\alpha_r) = 0$

and if we subtract (**) from (*) we have

$$[a_1 - \sigma_k(a_1)].\sigma_i(\alpha_1) + \cdots + [a_{r-1} - \sigma_k(a_{r-1})]\sigma_i(\alpha_{r-1}) = 0$$

which is a non-trivial solution to the system (') having fewer than r elements different from 0, contrary to the choice of r.

Corollary 1. If F is the fixed field for the finite group G, then each automorphism σ that leaves F fixed must belong to G.

(E/F) = order of G = n. Assume there is a σ not in G. Then F would remain fixed under the $n + 1$ elements consisting of σ and the elements of G, thus contradicting the corollary to Theorem 13.

Corollary 2. There are no two finite groups G_1 and G_2 with the same fixed field.

This follows immediately from Corollary 1.

If f(x) is a polynomial in F, then f(x) is called separable if its irreducible factors do not have repeated roots. If E is an extension of the field F, the element α of E is called separable if it is root of a separable polynomial f(x) in F, and E is called a separable extension if each element of E is separable.

THEOREM 15. E is a normal extension of F if and only if E is the splitting field of a separable polynomial p(x) in F.

Sufficiency. Under the assumption that E splits p(x) we prove that E is a normal extension of F.

If all roots of p(x) are in F, then our proposition is trivial, since then E = F and only the unit automorphism leaves F fixed.

Let us suppose p(x) has $n > 1$ roots in E but not in F. We make the inductive assumption that for all pairs of fields with fewer than n roots of p(x) outside of F our proposition holds.

Let $p(x) = p_1(x) \cdot p_2(x) \cdot \ldots \cdot p_r(x)$ be a factorization of p(x) into irreducible factors. We may suppose one of these to have a degree greater than one, for otherwise p(x) would split in F. Suppose deg $p_1(x) = s > 1$. Let α_1 be a root of $p_1(x)$. Then $(F(\alpha_1)/F) = \deg p_1(x) = s$. If we consider $F(\alpha_1)$ as the new ground field, fewer roots of p(x) than n are outside. From the fact that p(x) lies in $F(\alpha_1)$ and E is a splitting field of p(x) over $F(\alpha_1)$, it follows by our inductive assumption that E is a normal extension of $F(\alpha_1)$. Thus, each element in E which is not in $F(\alpha_1)$ is moved by at least one automorphism which leaves $F(\alpha_1)$ fixed.

p(x) being separable, the roots $\alpha_1, \alpha_2, \ldots, \alpha_s$ of $p_1(x)$ are a distinct elements of E. By Theorem 8 there exist isomorphisms

$\sigma_1, \sigma_2, \ldots, \sigma_s$ mapping $F(\alpha_1)$ on $F(\alpha_1), F(\alpha_2), \ldots, F(\alpha_s)$, respectively, which are each the identity on F and map α_1 on $\alpha_1, \alpha_2, \ldots, \alpha_s$ respectively. We now apply Theorem 10. E is a splitting field of $p(x)$ in $F(\alpha_1)$ and is also a splitting field of $p(x)$ in $F(\alpha_i)$. Hence, the isomorphism σ_i, which makes $p(x)$ in $F(\alpha_1)$ correspond to the same $p(x)$ in $F(\alpha_i)$, can be extended to an isomorphic mapping of E onto E, that is, to an automorphism of E that we denote again by σ_i. Hence, $\sigma_1, \sigma_2, \ldots, \sigma_s$ are automorphisms of E that leave F fixed and map α_1 onto $\alpha_1, \alpha_2, \ldots \alpha_n$.

Now let θ be an element that remains fixed under all automorphisms of E that leave F fixed. We know already that it is in $F(\alpha_1)$ and hence has the form

$$\theta = c_o + c_1\alpha_1 + c_2\alpha_1^2 + \ldots + c_{s-1}\alpha_1^{s-1}$$

where the c_i are in F. If we apply σ_i to this equation we get, since $\sigma_i(\theta) = \theta$:

$$\theta = c_o + c_1\alpha_i + c_2\alpha_i^2 + \ldots + c_{s-1}\alpha_i^{s-1}$$

The polynomial $c_{s-1}x^{s-1} + c_{s-2}x^{s-2} + \ldots + c_1x + (c_o - \theta)$ has therefore the s distinct roots $\alpha_1, \alpha_2, \ldots, \alpha_s$. These are more than its degree. So all coefficients of it must vanish, among them $c_o - \theta$. This shows θ in F.

Necessity. If E is a normal extension of F, then E is splitting field of a separable polynomial $p(x)$. We first prove the

Lemma. If E is a normal extension of F, then E is a separable extension of F. Moreover any element of E is a root of an equation over F which splits completely in E.

Let $\sigma_1, \sigma_2, \ldots, \sigma_s$ be the group G of automorphisms of E whose fixed field is F. Let a be an element of E, and let $a, a_2, a_3, \ldots, a_r$ be the set of distinct elements in the sequence $\sigma_1(a), \sigma_2(a), \ldots, \sigma_s(a)$. Since G is a group,

$$\sigma_j(a_i) = \sigma_j(\sigma_k(a)) = \sigma_j\sigma_k(a) = \sigma_m(a) = a_n.$$

Therefore, the elements $a, a_2, \ldots, a_r$ are permuted by the automorphisms of G. The coefficients of the polynomial $f(x) = (x-a)(x-a_2)\ldots(x-a_r)$ are left fixed by each automorphism of G, since in its factored form the factors of $f(x)$ are only permuted. Since the only elements of E which are left fixed by all the automorphisms of G belong to F, $f(x)$ is a polynomial in F. If $g(x)$ is a polynomial in F which also has a as root, then applying the automorphisms of G to the expression $g(a) = 0$ we obtain $g(a_i) = 0$, so that the degree of $g(x) \geq s$. Hence $f(x)$ is irreducible, and the lemma is established.

To complete the proof of the theorem, let $\omega_1, \omega_2, \ldots, \omega_t$ be a generating system for the vector space E over F. Let $f_i(x)$ be the separable polynomial having ω_i as a root. Then E is the splitting field of $p(x) = f_1(x) \cdot f_2(x) \cdot \ldots \cdot f_t(x)$.

If $f(x)$ is a polynomial in a field F, and E the splitting field of $f(x)$, then we shall call the group of automorphisms of E over F the group of the equation $f(x) = 0$. We come now to a theorem known in algebra as the Fundamental Theorem of Galois Theory which gives the relation between the structure of a splitting field and its group of automorphisms.

THEOREM 16. (Fundamental Theorem). If $p(x)$ is a separable polynomial in a field F, and G the group of the equation $p(x) = 0$ where E is the splitting field of p(x), then: (1) Each intermediate field, B,is the fixed field for a subgroup G_B of G, and distinct subgroups have distinct fixed fields. We say B and G_B "belong" to each other. (2) The intermediate field B is a normal extension of F if and only if the subgroup G_B is a normal subgroup of G. In this case the group of automorphisms of B which leaves F fixed is isomorphic to the factor group (G/G_B). (3) For each intermediate field B, we have (B/F) = index of G_B and (E/B) = order of G_B.

The first part of the theorem comes from the observation that E is splitting field for p(x) when p(x) is taken to be in any intermediate field. Hence, E is a normal extension of each intermediate field B, so that B is the fixed field of the subgroup of G consisting of the automorphisms which leave B fixed. That distinct subgroups have distinct fixed fields is stated in Corollary 2 to Theorem 14.

Let B be any intermediate field. Since B is the fixed field for the subgroup G_B of G, by Theorem 14 we have (E/B) = order of G_B. Let us call $o(G)$ the order of a group G and $i(G)$ its index. Then $o(G) = o(G_B) \cdot i(G_B)$. But $(E/F) = o(G)$, and $(E/F) = (E/B) \cdot (B/F)$ from which $(B/F) = i(G_B)$, which proves the third part of the theorem.

The number $i(G_B)$ is equal to the number of left cosets of G_B. The elements of G, being automorphisms of E, are isomorphisms of B; that is, they map B isomorphically into some other subfield of E and are the identity on F. The elements of G in any one coset of G_B map B in the same way. For let $\sigma \cdot \sigma_1$ and $\sigma \cdot \sigma_2$ be two elements of the coset σG_B. Since σ_1 and σ_2 leave B fixed, for each α in B we have $\sigma\sigma_1(a) = \sigma(a) = \sigma\sigma_2(a)$. Elements of different cosets give different isomorphisms, for if σ and τ give the same isomorphism, $\sigma(a) = \tau(a)$ for each a in B, then $\sigma^{-1}\tau(a) = a$ for each a in B. Hence, $\sigma^{-1}\tau = \sigma_1$, where σ_1 is an element of G_B. But then $\tau = \sigma\sigma_1$ and $\tau G_B = \sigma\sigma_1 G_B = \sigma G_B$ so that σ and τ belong to the same coset.

Each isomorphism of B which is the identity on F is given by an automorphism belonging to G. For let σ be an isomorphism mapping B on B' and the identity on F. Then under σ, $p(x)$ corresponds to $p(x)$, and E is the splitting field of $p(x)$ in B and of $p(x)$ in B'. By Theorem 10, σ can be extended to an automorphism σ' of E, and since σ' leaves F fixed it belongs to G. Therefore, the number of distinct isomorphisms of B is equal to the number of cosets of G_B and is therefore equal to (B/F).

The field σB onto which σ maps B has obviously $\sigma G_B \sigma^{-1}$ as corresponding group, since the elements of σB are left invariant by precisely this group.

If B is a normal extension of F, the number of distinct automorphisms of B which leave F fixed is (B/F) by Theorem 14. Conversely, if the number of automorphisms is (B/F) then B is a normal extension, because if F' is the fixed field of all these automorphisms, then $F \subset F' \subset B$, and by Theorem 14, (B/F') is equal to the number of automorphisms in the group, hence $(B/F') = (B/F)$. From $(B/F) = (B/F')(F'/F)$ we have $(F'/F) = 1$ or $F = F'$. Thus, B is a normal extension of F if and only if the number of automorphisms of B is (B/F).

B is a normal extension of F if and only if each isomorphism of B into E is an automorphism of B. This follows from the fact that each of the above conditions are equivalent to the assertion that there are the same number of isomorphisms and automorphisms. Since, for each σ, $B = \sigma B$ is equivalent to $\sigma G_B \sigma^{-1} \subset G_B$, we can finally say that B is a normal extension of F and only if G_B is a normal subgroup of G.

As we have shown, each isomorphism of B is described by the effect of the elements of some left coset of G_B. If B is a normal extension these isomorphisms are all automorphisms, but in this case the cosets are elements of the factor group (G/G_B). Thus, each automorphism of B corresponds uniquely to an element of (G/G_B) and conversely. Since multiplication in (G/G_B) is obtained by iterating the mappings, the correspondence is an isomorphism between (G/G_B) and the group of automorphisms of B which leave F fixed. This completes the proof of Theorem 16.

I. Finite Fields.

It is frequently necessary to know the nature of a finite subset of a field which under multiplication in the field is a group. The answer to this question is particularly simple.

THEOREM 17. If S is a finite subset ($\neq 0$) of a field F which is a group under multiplication in F, then S is a cyclic group.

The proof is based on the following lemmas for abelian groups.

Lemma 1. If in an abelian group A and B are two elements of orders a and b, and if c is the least common multiple of a and b, then there is an element C of order c in the group.

Proof: (a) If a and b are relatively prime, $C = AB$ has the required order ab. The order of $C^a = B^a$ is b and therefore c is divisible by b. Similarly it is divisible by a. Since $C^{ab} = 1$ it follows $c = ab$.

(b) If d is a divisor of a, we can find in the group an element of order d. Indeed $A^{a/d}$ is this element.

(c) Now let us consider the general case. Let $p_1, p_2, \ldots, p_r$ be the prime numbers dividing either a or b and let

$$a = p_1^{n_1} p_2^{n_2} \ldots p_r^{n_r}$$
$$b = p_1^{m_1} p_2^{m_2} \ldots p_r^{m_r}:$$

Call t_i the larger of the two numbers n_i and m_i. Then

$$c = p_1^{t_1} p_2^{t_2} \ldots p_r^{t_r}.$$

According to (b) we can find in the group an element of order $p_i^{n_i}$ and one of order $p_i^{m_i}$. Thus there is one of order $p_i^{t_i}$. Part (a) shows that the product of these elements will have the desired order c.

Lemma 2. If there is an element C in an abelian group whose order c is maximal (as is always the case if the group is finite) then c is divisible by the order a of every element A in the group; hence $x^c = 1$ is satisfied by each element in the group.

Proof: If a does not divide c, the greatest common multiple of a and c would be larger than c and we could find an element of that order, thus contradicting the choice of c.

We now prove Theorem 17. Let n be the order of S and r the largest order occuring in S. Then $x^r - 1 = 0$ is satisfied for all elements of S. Since this polynomial of degree r in the field cannot have more than r roots, it follows that $r \geq n$. On the other hand $r \leq n$ because the order of each element divides n. S is therefore a cyclic group consisting of $1, \epsilon, \epsilon^2, \ldots, \epsilon^{n-1}$ where $\epsilon^n = 1$.

Theorem 17 could also have been based on the decomposition theorem for abelian groups having a finite number of generators. Since this theorem will be needed later, we interpolate a proof of it here.

Let G be an abelian group, with group operation written as +. The element $g_1, \ldots, g_k$ will be said to generate G if each element g of G can be written as sum of multiples of $g_1, \ldots, g_k$, $g = n_1 g_1 + \ldots + n_k g_k$. If no set of fewer than k elements generate G, then $g_1, \ldots, g_k$ will be called a minimal generating system. Any group having a finite generating system admits a minimal generating system. In particular, a finite group always admits a minimal generating system.

From the identity $n_1(g_1 + mg_2) + (n_2 - n_1 m)g_2 = n_1 g_1 + n_2 g_2$ it follows that if $g_1, g_2, \ldots, g_k$ generate G, also $g_1 + mg_2$, $g_2, \ldots, g_k$ generate G.

An equation $m_1 g_1 + m_2 g_2 + \ldots + m_k g_k = 0$ will be called a relation between the generators, and $m_1, \ldots, m_k$ will be called the coefficients in the relation.

We shall say that the abelian group G is the direct product of its subgroups $G_1, G_2, \ldots, G_k$ if each $g \epsilon G$ is uniquely representable as a sum $g = x_1 + x_2 + \ldots + x_k$, where $x_i \epsilon G_i$, $i = 1, \ldots, k$.

Decomposition Theorem. Each abelian group having a finite number of generators is the direct product of cyclic subgroups $G_1, \ldots, G_n$ where the order of G_i divides the order of G_{i+1}, $i = 1, \ldots, n\text{-}1$ and n is the number of elements in a minimal generating system. ($G_r, G_{r+1}, \ldots, G_n$ may each be infinite, in which case, to be precise, $0(G_i) | 0(G_{i+1})$ for $i = 1, 2, \ldots, r\text{-}2$).

We assume the theorem true for all groups having minimal generating systems of k-1 elements. If $n = 1$ the group is cyclic and the theorem trivial. Now suppose G is an abelian group having a minimal generating system of k elements. If no minimal generating system satisfies a non-trivial relation, then let $g_1, g_2, \ldots, g_k$ be a minimal generating system and $G_1, G_2, \ldots, G_k$ be the cyclic groups generated by them. For each $g \in G$, $g = n_1 g_1 + \ldots + n_k g_k$ where the expression is unique; otherwise we should obtain a relation. Thus the theorem would be true. Assume now that some non-trivial relations hold for some minimal generating systems. Among all relations between minimal generating systems, let

$$(1) \qquad m_1 g_1 + \ldots + m_k g_k = 0$$

be a relation in which the smallest positive coefficient occurs. After an eventual reordering of the generators we can suppose this coefficient to be m_1. In any other relation between $g_1, \ldots, g_k$.

$$(2) \qquad n_1 g_1 + \ldots + n_k g_k = 0$$

we must have m_1/n_1. Otherwise $n_1 = qm_1 + r$, $0 < r < m_1$ and q times relation (1) subtracted from relation (2) would yield a relation with a coefficient $r < m_1$. Also in relation (1) we must have m_1/m_i, $i = 2, \ldots, k$.

For suppose m_1 does not divide one coefficient, say m_2. Then $m_2 = qm_1 + r$, $0 < r < m_1$. In the generating system $g_1 + g_2, g_2, \ldots, g_k$ we should have a relation $m_1(g_1 + qg_2) + rg_2 + m_3g_3 + \ldots + m_kq_k = 0$ where the coefficient r contradicts the choice of m_1. Hence $m_2 = q_2m_1$, $m_3 = q_3m_1, \ldots, m_k = q_km_1$. The system $\bar{g}_1 = g_1 + q_2g_2 + \ldots + q_kg_k, g_2, \ldots, g_k$ is minimal generating, and $m_1\bar{g}_1 = 0$. In any relation $0 = n_1\bar{g}_1 + n_2g_2 + \ldots + n_kg_k$ since m_1 is a coefficient in a relation between $\bar{g}_1, g_2, \ldots, g_k$ our previous argument yields $m_1 \mid n_1$, and hence $n_1\bar{g}_1 = 0$.

Let G' be the subgroup of G generated by $g_2, \ldots, g_k$ and G_1 the cyclic group of order m_1 generated by $\bar{g}_1$. Then G is the direct product of G_1 and G'. Each element g of G can be written

$$g = n_1\bar{g}_1 + n_2g_2 + \ldots + n_kg_k = n_1\bar{g}_1 + g'.$$

The representation is unique, since $n_1\bar{g}_1 + g' = n_1'\bar{g}_1 + g''$ implies the relation $(n_1 - n_1')\bar{g}_1 + (g' - g'') = 0$, hence $(n_1 - n_1')\bar{g}_1 = 0$, so that $n_1\bar{g}_1 = n_1'\bar{g}_1$ and also $g' = g''$.

By our inductive hypothesis, G' is the direct product of k-1 cyclic groups generated by elements $\bar{g}_2, \bar{g}_3, \ldots, \bar{g}_k$ whose respective orders $t_2, \ldots, t_k$ satisfy $t_i \mid t_{i+1}$, $i = 2, \ldots, k-1$. The preceding argument applied to the generators $\bar{g}_1, \bar{g}_2, \ldots, \bar{g}_k$ yields $m_1 \mid t_2$, from which the theorem follows.

By a <u>finite field</u> is meant one having only a finite number of elements.

<u>Corollary.</u> <u>The non-zero elements of a finite field form a cyclic group.</u>

If a is an element of a field F, let us denote the n-fold of a, i.e., the element of F obtained by adding a to itself n times, by na. It is obvious that $n \cdot (m \cdot a) = (nm) \cdot a$ and $(n \cdot a)(m \cdot b) = nm \cdot ab$. If for one element $a \neq 0$, there is an integer n such that $n \cdot a = 0$ then $n \cdot b = 0$ for each b in F, since $n \cdot b = (n \cdot a)(a^{-1} b) = 0 \cdot a^{-1} b = 0$. If there is a positive integer p such that $p \cdot a = 0$ for each a in F, and if p is the smallest integer with this property, then F is said to have the <u>characteristic</u> p. If no such positive integer exists then we say F has characteristic 0. <u>The characteristic of a field is always a prime number</u>, for if $p = r \cdot s$ then $pa = rs \cdot a = r \cdot (s \cdot a)$. However, $s \cdot a = b \neq 0$ if $a \neq 0$ and $r \cdot b \neq 0$ since both r and s are less than p, so that $pa \neq 0$ contrary to the definition of the characteristic. If $na = 0$ for $a \neq 0$, then p divides n, for $n = qp + r$ where $0 \leq r < p$ and $na = (qp + r)a = qpa + ra$. Hence $na = 0$ implies $ra = 0$ and from the definition of the characteristic since $r < p$, we must have $r = 0$.

If F is a finite field having q elements and E an extension of F such that $(E/F) = n$, then E has q^n elements. For if $\omega_1, \omega_2, \ldots, \omega_n$ is a basis of E over F, each element of E can be uniquely represented as a linear combination $x_1\omega_1 + x_2\omega_2 + \ldots + x_n\omega_n$ where the x_i belong to F. Since each x_i can assume q values in F, there are q^n distinct possible choices of $x_1, \ldots, x_n$ and hence q^n distinct elements of E. E is finite, hence, <u>there is an element α of E so that $E = F(\alpha)$.</u> (The non-zero elements of E form a cyclic group generated by α).

If we denote by $P \equiv [0, 1, 2, \ldots, p\text{-}1]$ the set of multiples of the unit element in a field F of characteristic p, then P is a subfield of F having p distinct elements. In fact, P is isomorphic to the field of integers reduced mod p. If F is a finite field, then the degree of F over

P is finite, say $(F/P) = n$, and F contains p^n elements. In other words, <u>the order of any finite field is a power of its characteristic.</u>

If F and F' are two finite fields having the same order q, then by the preceding, they have the same characteristic since q is a power of the characteristic. The multiples of the unit in F and F' form two fields P and P' which are isomorphic.

The non-zero elements of F and F' form a group of order q-1 and, therefore, satisfy the equation $x^{q-1} - 1 = 0$. The fields F and F' are splitting fields of the equation $x^{q-1} = 1$ considered as lying in P and P' respectively. By Theorem 10, the isomorphism between P and P' can be extended to an isomorphism between F and F'. We have thus proved

<u>THEOREM 18.</u> <u>Two finite fields having the same number of elements are isomorphic.</u>

<u>Differentiation.</u> If $f(x) = a_o + a_1x + \ldots + a_nx^n$ is a polynomial in a field F, then we define $f' = a_1 + 2a_2x + \ldots + na_nx^{n-1}$. The reader may readily verify that for each pair of polynomials f and g we have

$$(f + g)' = f' + g'$$

$$(f\ g)' = fg' + gf'$$

$$(f^n)' = nf^{n-1} \cdot f'$$

<u>THEOREM 19.</u> <u>The polynomial f has repeated roots if and only if in the splitting field E the polynomials f and f' have a common root. This condition is equivalent to the assertion that f and f' have a</u>

common factor of degree greater than 0 in F.

If α is a root of multiplicity k of $f(x)$ then $f = (x-\alpha)^k Q(x)$ where $Q(\alpha) \neq 0$. This gives

$$f' = (x-\alpha)^k Q'(x) + k(x-\alpha)^{k-1} Q(x) = (x-\alpha)^{k-1}[(x-\alpha)Q'(x) + kQ(x)].$$

If $k > 1$, then α is a root of f' of multiplicity at least $k-1$. If $k = 1$, then $f'(x) = Q(x) + (x-\alpha)Q'(x)$ and $f'(\alpha) = Q(\alpha) \neq 0$. Thus, f and f' have a root α in common if and only if α is a root of f of multiplicity greater than 1.

If f and f' have a root α in common then the irreducible polynomial in F having α as root divides both f and f'. Conversely, any root of a factor common to both f and f' is a root of f and f'.

Corollary. If F is a field of characteristic 0 then each irreducible polynomial in F is separable.

Suppose to the contrary that the irreducible polynomial $f(x)$ has a root α of multiplicity greater than 1. Then, $f'(x)$ is a polynomial which is not identically zero (its leading coefficient is a multiple of the leading coefficient of $f(x)$ and is not zero since the characteristic is 0) and of degree 1 less than the degree of $f(x)$. But α is also a root of $f'(x)$ which contradicts the irreducibility of $f(x)$.

J. Roots of Unity.

If F is a field having any characteristic p, and E the splitting field of the polynomial $x^n - 1$ where p does not divide n, then we shall refer to E as the field generated out of F by the adjunction of a primitive n^{th} root of unity.

The polynomial $x^n - 1$ does not have repeated roots in E, since its derivative, nx^{n-1}, has only the root 0 and has, therefore, no roots in common with $x^n - 1$. Thus, E is a normal extension of F. If $\epsilon_1, \epsilon_2, \ldots, \epsilon_n$ are the roots of $x^n - 1$ in E, they form a group under multiplication and by Theorem 17 this group will be cyclic. If $1, \epsilon, \epsilon^2, \ldots, \epsilon^{n-1}$ are the elements of the group, we shall call ϵ a primitive n^{th} root of unity. The smallest power of ϵ which is 1 is the nth.

THEOREM 20. If E is the field generated from F by a primitive n^{th} root of unity, then the group G of E over F is abelian for any n and cyclic if n is a prime number.

We have $E = F(\epsilon)$, since the roots of $x^n - 1$ are powers of ϵ. Thus, if σ and τ are distinct elements of G, $\sigma(\epsilon) \neq \tau(\epsilon)$. But $\sigma(\epsilon)$ is a root of $x^n - 1$ and, hence, a power of ϵ. Thus, $\sigma(\epsilon) = \epsilon^{n_\sigma}$ where n_σ is an integer $1 \leq n_\sigma < n$. Moreover, $\tau\sigma(\epsilon) = \tau(\epsilon^{n_\sigma}) = (\tau(\epsilon))^{n_\sigma} = \epsilon^{n_\tau \cdot n_\sigma} = \sigma\tau(\epsilon)$. Thus, $n_{\sigma\tau} = n_\sigma n_\tau$ mod n. Thus, the mapping of σ on n_σ is a homomorphism of G into a multiplicative subgroup of the integers mod n. Since $\tau \neq \sigma$ implies $\tau(\epsilon) \neq \sigma(\epsilon)$, it follows that $\tau \neq \sigma$ implies $n_\sigma \neq n_\tau$ mod n. Hence, the homomorphism is an isomorphism. If n is a prime number, the multiplicative group of numbers forms a cyclic group.

K. Noether Equations.

If E is a field, and $G = (\sigma, \tau, \ldots)$ a group of automorphisms of E, any set of elements $x_\sigma, x_\tau, \ldots$ in E will be said to provide a solution to Noether's equations if $x_\sigma \cdot \sigma(x_\tau) = x_{\sigma\tau}$ for each σ and τ in G. If one element $x_\sigma = 0$ then $x_\tau = 0$ for each $\tau \in G$. As τ traces G, $\sigma\tau$ assumes all values in G, and in the above equation $x_{\sigma\tau} = 0$ when $x_\sigma = 0$. Thus, in any solution of the Noether equations no element $x_\sigma = 0$ unless the solution is completely trivial. We shall assume in the sequel that the trivial solution has been excluded.

THEOREM 21. The system $x_\sigma, x_\tau, \ldots$ is a solution to Noether's equations if and only if there exists an element α in E, such that $x_\sigma = \alpha/\sigma(\alpha)$ for each σ.

For any α, it is clear that $x_\sigma = \alpha/\sigma(\alpha)$ is a solution to the equations, since

$$\alpha/\sigma(\alpha)\cdot\sigma(\alpha/\tau(\alpha)) = \alpha/\sigma(\alpha)\cdot\sigma(\alpha)/\sigma\tau(\alpha) = \alpha/\sigma\tau(\alpha).$$

Conversely, let $x_\sigma, x_\tau, \ldots$ be a non-trivial solution. Since the automorphisms $\sigma, \tau, \ldots$ are distinct they are linearly independent, and the equation $x_\sigma\cdot\sigma(z) + x_\tau\tau(z) + \ldots = 0$ does not hold identically. Hence, there is an element a in E such that $x_\sigma\sigma(a) + x_\tau\tau(a) + \ldots = \alpha \neq 0$. Applying σ to α gives

$$\sigma(\alpha) = \sum_{\tau\epsilon G} \sigma(x_\tau)\cdot\sigma\tau(a).$$

Multiplying by x_σ gives

$$x_\sigma\cdot\sigma(\alpha) = \sum_{\tau\epsilon G} x_\sigma\sigma(x_\tau)\cdot\sigma\tau(a).$$

Replacing $x_\sigma\cdot\sigma(x_\tau)$ by $x_{\sigma\tau}$ and noting that $\sigma\tau$ assumes all values in G when τ does, we have

$$x_\sigma\cdot\sigma(\alpha) = \sum_{\tau\epsilon G} x_t\tau(a) = \alpha$$

so that

$$x_\sigma = \alpha/\sigma(\alpha).$$

A solution to the Noether equations defines a mapping C of G into E, namely, $C(\sigma) = x_\sigma$. If F is the fixed field of G, and the elements x_σ lie in F, then C is a character of G. For $C(\sigma\tau) = x_{\sigma\tau} = x_\sigma\cdot\sigma(x_\tau) = x_\sigma x_\tau = C(\sigma)\cdot C(\tau)$ since $\sigma(x_\tau) = x_t$ if $x_\tau \epsilon F$. Conversely, each character C of G in F provides a solution

to the Noether equations. Call $C(\sigma) = x_\sigma$. Then, since $x_\tau \in F$, we have $\sigma(x_\tau) = x_\tau$. Thus,
$x_\sigma \cdot \sigma(x_\tau) = x_\sigma \cdot x_\tau = C(\sigma) \cdot C(\tau) = C(\sigma\tau) = x_{\sigma\tau}$. We therefore have, by combining this with Theorem 21,

THEOREM 22. If G is the group of the normal field E over F, then for each character C of G into F there exists an element α in E such that $C(\sigma) = \alpha/\sigma(\alpha)$ and, conversely, if $\alpha/\sigma(\alpha)$ is in F for each σ, then $C(\sigma) = \alpha/\sigma(\alpha)$ is a character of G. If r is the least common multiple of the orders of elements of G, then $\alpha^r \in F$.

We have already shown all but the last sentence of Theorem 22. To prove this we need only show $\sigma(\alpha^r) = \alpha^r$ for each $\sigma \in G$. But $\alpha^r/\sigma(\alpha^r) = (\alpha/\sigma(\alpha))^r = (C(\sigma))^r = C(\sigma^r) = C(I) = 1$.

L. Kummer's Fields.

If F contains a primitive n^{th} root of unity, any splitting field E of a polynomial $(x^n - a_1)(x^n - a_2)\ldots(x^n - a_r)$ where $a_i \in F$ for $i = 1, 2, \ldots, r$ will be called a Kummer extension of F, or more briefly, a Kummer field.

If a field F contains a primitive n^{th} root of unity, the number n is not divisible by the characteristic of F. Suppose, to the contrary, F has characteristic p and $n = qp$. Then $y^p - 1 = (y - 1)^p$ since in the expansion of $(y - 1)^p$ each coefficient other than the first and last is divisible by p and therefore is a multiple of the p-fold of the unit of F and thus is equal to 0. Therefore $x^n - 1 = (x^q)^p - 1 = (x^q - 1)^p$ and $x^n - 1$ cannot have more than q distinct roots. But we assumed that F has a primitive n^{th} root of unity and $1, \epsilon, \epsilon^2, \ldots, \epsilon^{n-1}$ would be n distinct roots of $x^n - 1$. It follows that n is not divisible by the characteristic of F. For a Kummer field E, none of the factors $x^n - a_i$, $a_i \neq 0$ has repeated roots since the derivative, nx^{n-1}, has only the root 0 and has therefore no roots in common with $x^n - a_i$. Therefore, the irreducible factors of $x^n - a_i$ are separable, so that E is a normal extension of F.

Let α_i be a root of $x^n - a_i$ in E. If $\epsilon_1, \epsilon_2, \ldots, \epsilon_n$ are the n distinct n^{th} roots of unity in F, then $\alpha_i\epsilon_1, \alpha_i\epsilon_2, \ldots, \alpha_i\epsilon_n$ will be n distinct roots of $x^n - a_i$, and hence will be the roots of $x^n - a_i$, so that $E = F(\alpha_1, \alpha_2, \ldots, \alpha_r)$. Let σ and τ be two automorphisms in the group G of E over F. For each α_i, both σ and τ map α_i on some other root of $x^n - a_i$. Thus $\tau(\alpha_i) = \epsilon_{i\tau}\alpha_i$ and $\sigma(\alpha_i) = \epsilon_{i\sigma}\alpha_i$ where $\epsilon_{i\sigma}$ and $\epsilon_{i\tau}$ are n^{th} roots of unity in the basic field F. It follows that $\tau(\sigma(\alpha_i)) = \tau(\epsilon_{i\sigma}\alpha_i) = \epsilon_{i\sigma}\tau(\alpha_i) = \epsilon_{i\sigma}\epsilon_{i\tau}\alpha_i = \sigma(\tau(\alpha_i)$. Since σ and τ are commutative over the generators of E, they commute over each element of E. Hence, G is commutative. If $\sigma \in G$, then $\sigma(\alpha_i) = \epsilon_{i\sigma}\alpha_i$, $\sigma^2(\alpha_i) = \epsilon_{i\sigma}{}^2\alpha_i$, etc. Thus, $\sigma^{n_i}(\alpha_i) = \alpha_i$ for n_i such that $\epsilon_{i\sigma}{}^{n_i} = 1$. Since the order of an nth root of unity is a divisor of n, we have n_i a divisor of n and the least common multiple m of $n_1, n_2, \ldots, n_r$ is a divisor of n. Since $\sigma^m(\alpha_i) = \alpha_i$ for $i = 1, 2, \ldots, r$ it follows that m is the order of σ. Hence, the order of each element of G is a divisor of n and, therefore, the least common multiple r of the orders of the elements of G is a divisor of n. If ϵ is a primitive n^{th} root of unity, then $\epsilon^{n/r}$ is a primitive r^{th} root of unity. These remarks can be summarized in the following.

THEOREM 23. If E is a Kummer field, i.e., a splitting field of $p(x) = (x^n - a_1)(x^n - a_2)\ldots(x^n - a_t)$ where a_i lie in F, and F contains a primitive n^{th} root of unity, then: (a) E is a normal extension of F; (b) the group G of E over F is abelian, (c) the least common multiple of the orders of the elements of G is a divisor of n.

Corollary. If E is the splitting field of $x^p - a$, and F contains a primitive p^{th} root of unity where p is a prime number, then either E = F and $x^p - a$ is split in F, or $x^p - a$ is irreducible and the group of E over F is cyclic of order p.

The order of each element of G is, by Theorem 23, a divisor of p and, hence, if the element is not the unit its order must be p. If α is a root of $x^p - a$, then $\alpha, \epsilon\alpha, \ldots, \epsilon^{p-1}\alpha$ are all the roots of $x^p - a$ so that $F(\alpha) = E$ and $(E/F) \leq p$. Hence, the order of G does not exceed p so that if G has one element different from the unit, it and its powers must constitute all of G. Since G has p distinct elements and their behavior is determined by their effect on α, then α must have p distinct images. Hence, the irreducible equation in F for α must be of degree p and is therefore $x^p - a = 0$.

The properties (a), (b) and (c) in Theorem 23 actually characterize Kummer fields.

Let us suppose that E is a normal extension of a field F, whose group G over F is abelian. Let us further assume that F contains a primitive r^{th} root of unity where r is the least common multiple of the orders of elements of G.

The group of characters X of G into the group of rth roots of unity is isomorphic to G. Moreover, to each $\sigma \in G$, if $\sigma \neq 1$, there exists a character $C \in X$ such that $C(\sigma) \neq 1$. Write G as the direct product of the cyclic groups $G_1, G_2, \ldots, G_t$ of orders $m_1 \mid m_2 \mid \ldots \mid m_t$. Each $\sigma \in G$ may be written $\sigma = \sigma_1^{\nu_1}\sigma_2^{\nu_2}\ldots\sigma_t^{\nu_t}$. Call C_i the character sending σ_i into ϵ_i, a primitive m_i^{th} root of unity and σ_j into 1 for $j \neq i$. Let C be any character. $C(\sigma_i) = \epsilon_i^{\mu_i}$, then we have $C = C_1^{\mu_i} \cdot C_2^{\mu_i} \ldots C_t^{\mu_i}$. Conversely, $C_1^{\mu_1} \ldots C_t^{\mu_t}$ defines a character. Since the order of C_i is m_i, the character group X of G is isomorphic to G. If $\sigma \neq 1$, then in $\sigma = \sigma_1^{\nu_1}\sigma_2^{\nu_2}\ldots\sigma_t^{\nu_t}$ at least one ν_i, say ν_1, is not divisible by m_1. Thus $C_1(\sigma) = \epsilon_1^{\nu_1} \neq 1$.

Let A denote the set of those non-zero elements α of E for which $\alpha^r \in F$ and let F_1 denote the non-zero elements of F. It is obvious that A is a multiplicative group and that F_1 is a subgroup of A. Let A^r denote the set of rth powers of elements in A and F_1^r the set of rth powers of elements of F_1. The following theorem provides in most applications a convenient method for computing the group G.

THEOREM 24. The factor groups (A/F_1) and (A^r/F_1^r) are isomorphic to each other and to the groups G and X.

We map A on A^r by making $\alpha \in A$ correspond to $\alpha^r \in A^r$. If $a^r \in F_1^r$, where $a \in F_1$ then $b \in A$ is mapped on a^r if and only if $b^r = a^r$, that is, if b is a solution to the equation $x^r - a^r = 0$. But $a, \epsilon a, \epsilon^2 a, \ldots, \epsilon^{r-1} a$ are distinct solutions to this equation and since ϵ and a belong to F_1, it follows that b must be one of these elements and must belong to F_1. Thus, the inverse set in A of the subgroup F_1^r of A^r is F_1, so that the factor groups (A/F_1) and (A^r/F_1^r) are isomorphic.

If α is an element of A, then $(\alpha/\sigma(\alpha))^r = \alpha^r/\sigma(\alpha^r) = 1$. Hence, $\alpha/\sigma(\alpha)$ is an rth root of unity and lies in F_1. By Theorem 22, $\alpha/\sigma(\alpha)$ defines a character $C(\sigma)$ of G in F. We map α on the corresponding character C. Each character C is by Theorem 22, image of some α. Moreover, $\alpha \cdot \alpha'$ is mapped on the character $C^*(\sigma) = \alpha \cdot \alpha'/\sigma(\alpha \cdot \alpha') = \alpha \cdot \alpha'/\sigma(\alpha) \cdot \sigma(\alpha') = C(\sigma) \cdot C'(\sigma) = C \cdot C'(\sigma)$, so that the mapping is homomorphism. The kernel of this homomorphism is the set of those elements α for which $\alpha/\sigma(\alpha) = 1$ for each σ, hence is F_1. It follows, therefore, that (A/F_1) is isomorphic to X and hence also to G. In particular, (A/F_1) is a finite group.

We now prove the equivalence between Kummer fields and fields satisfying (a), (b) and (c) of Theorem 23.

THEOREM 25. If E is an extension field over F, then E is a Kummer field if and only if E is normal, its group G is abelian and F contains a primitive r^{th} root ϵ of unity where r is the least common multiple of the orders of the elements of G.

The necessity is already contained in Theorem 23. We prove the sufficiency. Out of the group A, let $\alpha_1 F_1, \alpha_2 F_1, \ldots, \alpha_t F_1$ be the cosets of F_1. Since $\alpha_i \in A$, we have $\alpha_i^r = a_i \in F$. Thus, α_i is a root of the equation $x^r - a_i = 0$ and since $\epsilon\alpha_i, \epsilon^2\alpha_i, \ldots, \epsilon^{r-1}\alpha_i$ are also roots, $x^r - a_i$ must split in E. We prove that E is the splitting field of $(x^r - a_1)(x^r - a_2)\ldots(x^r - a_t)$ which will complete the proof of the theorem. To this end it suffices to show that $F(\alpha_1, \alpha_2, \ldots, \alpha_t) = E$.

Suppose that $F(\alpha_1, \alpha_2, \ldots, \alpha_t) \neq E$. Then $F(\alpha_1, \ldots, \alpha_t)$ is an intermediate field between F and E, and since E is normal over $F(\alpha_1, \ldots, \alpha_t)$ there exists an automorphism $\sigma \in G$, $\sigma \neq 1$, which leaves $F(\alpha_1, \ldots, \alpha_t)$ fixed. There exists a character C of G for which $C(\sigma) \neq 1$. Finally, there exists an element α in E such that $C(\sigma) = \alpha/\sigma(\alpha) \neq 1$. But $\alpha^r \in F_1$ by Theorem 22, hence $\alpha \in A$. Moreover, $A \subset F(\alpha_1, \ldots, \alpha_t)$ since all the cosets $\alpha_i F_1$ are contained in $F(\alpha_1, \ldots, \alpha_t)$. Since $F(\alpha_1, \ldots, \alpha_t)$ is by assumption left fixed by σ, $\sigma(\alpha) = \alpha$ which contradicts $\alpha/\sigma(\alpha) \neq 1$. It follows, therefore, that $F(\alpha_1, \ldots, \alpha_t) = E$.

Corollary. If E is a normal extension of F, of prime order p, and if F contains a primitive p^{th} root of unity, then E is splitting field of an irreducible polynomial $x^p - a$ in F.

E is generated by elements $\alpha_1, \ldots, \alpha_n$ where $\alpha_i^p \in F$. Let α_1 be not in F. Then $x^p - a$ is irreducible, for otherwise $F(\alpha_1)$ would be an intermediate field between F and E of degree less than p, and by the product theorem for the degrees, p would not be a prime number, contrary to assumption. $E = F(\alpha_1)$ is the splitting field of $x^p - a$.

M. Simple Extensions.

We consider the question of determining under what conditions an extension field is generated by a single element, called a primitive. We prove the following

THEOREM 26. A finite extension E of F is primitive over F if

<u>and only if there are only a finite number of intermediate fields.</u>

(a) Let $E = F(\alpha)$ and call $f(x) = 0$ the irreducible equation for α in F. Let B be an intermediate field and $g(x)$ the irreducible equation for α in B. The coefficients of $g(x)$ adjoined to F will generate a field B' between F and B. $g(x)$ is irreducible in B, hence also in B'. Since $E = B'(\alpha)$ we see $(E/B) = (E/B')$. This proves $B' = B$. So B is uniquely determined by the polynomial $g(x)$. But $g(x)$ is a divisor of $f(x)$, and there are only a finite number of possible divisors of $f(x)$ in E. Hence there are only a finite number of possible B's.

(b) Assume there are only a finite number of fields between E and F. Should F consist only of a finite number of elements, then E is generated by one element according to the Corollary on page 53. We may therefore assume F to contain an infinity of elements. We prove: To any two elements α, β there is a γ in E such that $F(\alpha, \beta) = F(\gamma)$. Let $\gamma = \alpha + a\beta$ with a in F but for the moment undetermined. Consider all the fields $F(\gamma)$ obtained in this way. Since we have an infinity of a's at our disposal, we can find two, say a_1 and a_2, such that the corresponding γ's, $\gamma_1 = \alpha + a_1\beta$ and $\gamma_2 = \alpha + a_2\beta$, yield the same field $F(\gamma_1) = F(\gamma_2)$. Since both γ_1 and γ_2 are in $F(\gamma_1)$, their difference (and therefore β) is in this field. Consequently also $\gamma_1 - a_1\beta = \alpha$. So $F(\alpha, \beta) \subset F(\gamma_1)$. Since $F(\gamma_1) \subset F(\alpha, \beta)$ our contention is proved. Select now η in E in such a way that $(F(\eta)/F)$ is as large as possible. Every element ϵ of E must be in $F(\eta)$ or else we could find an element δ such that $F(\delta)$ contains both η and ϵ. This proves $E = F(\eta)$.

THEOREM 27. If $E = F(\alpha_1, \alpha_2, \ldots, \alpha_n)$ is a finite extension of the field F, and $\alpha_1, \alpha_2, \ldots, \alpha_n$ are separable elements in E, then there exists a primitive θ in E such that $E = F(\theta)$.

Proof: Let $f_i(x)$ be the irreducible equation of α_i in F and let B be an extension of E that splits $f_1(x)f_2(x)\ldots f_n(x)$. Then B is normal over F and contains, therefore, only a finite number of intermediate fields (as many as there are subgroups of G). So the subfield E contains only a finite number of intermediate fields. Theorem 26 now completes the proof.

N. Existence of a Normal Basis.

The following theorem is true for any field though we prove it only in the case that F contains an infinity of elements.

THEOREM 28. If E is a normal extension of F and $\sigma_1, \sigma_2, \ldots, \sigma_n$ are the elements of its group G, there is an element θ in E such that the n elements $\sigma_1(\theta), \sigma_2(\theta), \ldots, \sigma_n(\theta)$ are linearly independent with respect to F.

According to Theorem 27 there is an α such that $E = F(\alpha)$. Let $f(x)$ be the equation for α, put $\sigma_i(\alpha) = \alpha_i$,

$$g(x) = \frac{f(x)}{(x-\alpha)f'(\alpha)} \text{ and } g_i(x) = \sigma_i(g(x)) = \frac{f(x)}{(x-\alpha_i)f'(\alpha_i)}$$

$g_i(x)$ is a polynomial in E having α_k as root for $k \neq i$ and hence

$$(1) \qquad g_i(x)g_k(x) \equiv 0 \pmod{f(x)} \text{ for } i \neq k.$$

In the equation

$$(2) \qquad g_1(x) + g_2(x) + \ldots + g_n(x) - 1 = 0$$

the left side is of degree at most $n - 1$. If (2) is true for n different values of x, the left side must be identically 0. Such n values are

$\alpha_1, \alpha_2, \ldots, \alpha_n$, since $g_i(\alpha_i) = 1$ and $g_k(\alpha_i) = 0$ for $k \neq i$.

Multiplying (2) by $g_i(x)$ and using (1) shows:

(3) $$(g_i(x))^2 \equiv g_i(x) \pmod{f(x)}.$$

We next compute the determinant

(4) $$D(x) = |\sigma_i\sigma_k(g(x))| \quad i, k = 1, 2, \ldots, n$$

and prove $D(x) \neq 0$. If we square it by multiplying column by column and compute its value (mod $f(x)$) we get from (1), (2), (3) a determinant that has 1 in the diagonal and 0 elsewhere.

So

$$(D(x))^2 \equiv 1 \pmod{f(x)}.$$

$D(x)$ can have only a finite number of roots in F. Avoiding them we can find a value a for x such that $D(a) \neq 0$. Now set $\theta = g(a)$. Then the determinant

(5) $$|\sigma_i\sigma_k(\theta)| \neq 0.$$

Consider any linear relation

$x_1\sigma_1(\theta) + x_2\sigma_2(\theta) + \ldots + x_n\sigma_n(\theta) = 0$ where the x_i are in F. Applying the automorphism σ_i to it would lead to n homogeneous equations for the n unknowns x_i. (5) shows that $x_i = 0$ and our theorem is proved.

O. Theorem on Natural Irrationalities.

Let F be a field, $p(x)$ a polynomial in F whose irreducible factors are separable, and let E be a splitting field for $p(x)$. Let B be an arbitrary extension of F, and let us denote by EB the splitting field of $p(x)$ when $p(x)$ is taken to lie in B. If $\alpha_1, \ldots, \alpha_s$ are the roots of $p(x)$ in EB, then $F(\alpha_1, \ldots, \alpha_s)$ is a subfield of EB which is readily seen to form a splitting field for $p(x)$ in F. By Theorem 10, E and $F(\alpha_1, \ldots, \alpha_s)$

are isomorphic. There is therefore no loss of generality if in the sequel we take $E = F(\alpha_1, \ldots, \alpha_s)$ and assume therefore that E is a subfield of EB. Also, $EB = B(\alpha_1, \ldots, \alpha_s)$.

Let us denote by $E \cap B$ the intersection of E and B. It is readily seen that $E \cap B$ is a field and is intermediate to F and E.

THEOREM 29. If G is the group of automorphisms of E over F, and H the group of EB over B, then H is isomorphic to the subgroup of G having $E \cap B$ as its fixed field.

Each automorphism of EB over B simply permutes $\alpha_1, \ldots, \alpha_s$ in some fashion and leaves B, and hence also F, fixed. Since the elements of EB are quotients of polynomial expressions in $\alpha_1, \ldots, \alpha_s$ with coefficients in B, the automorphism is completely determined by the permutation it effects on $\alpha_1, \ldots, \alpha_s$. Thus, each automorphism of EB over B defines an automorphism of $E = F(\alpha_1, \ldots, \alpha_s)$ which leaves F fixed. Distinct automorphisms, since $\alpha_1, \ldots, \alpha_s$ belong to E, have different effects on E. Thus, the group H of EB over B can be considered as a subgroup of the group G of E over F. Each element of H leaves $E \cap B$ fixed since it leaves even all of B fixed. However, any element of E which is not in $E \cap B$ is not in B, and hence would be moved by at least one automorphism of H. It follows that $E \cap B$ is the fixed field of H.

Corollary. If, under the conditions of Theorem 29, the group G is of prime order, then either H = G or H consists of the unit element alone.

III APPLICATIONS

by

A. N. Milgram

A. Solvable Groups.

Before proceeding with the applications we must discuss certain questions in the theory of groups. We shall assume several simple propositions: (a) If N is a normal subgroup of the group G, then the mapping $f(x) = xN$ is a homomorphism of G on the factor group G/N. f is called the natural homomorphism. (b) The image and the inverse image of a normal subgroup under a homomorphism is a normal subgroup. (c) If f is a homomorphism of the group G on G', then setting $N' = f(N)$, and defining the mapping g as $g(xN) = f(x)N'$, we readily see that g is a homomorphism of the factor group G/N on the factor group G'/N'. Indeed, if N is the inverse image of N' then g is an isomorphism.

We now prove

THEOREM 1. (Zassenhaus). If U and V are subgroups of G, u and v normal subgroups of U and V, respectively, then the following three factor groups are isomorphic: $u(U\cap V)/u(U\cap v)$, $v(U\cap V)/v(u\cap V)$, $(U\cap V)/(u\cap V)(v\cap U)$.

It is obvious that $U \cap v$ is a normal subgroup of $U \cap V$. Let f be the natural mapping of U on U/u. Call $f(U\cap V) = H$ and $f(U\cap v) = K$. Then $f^{-1}(H) = u(U\cap V)$ and $f^{-1}(K) = u(U\cap v)$ from which it follows that $u(U\cap V)/u(U\cap v)$ is isomorphic to H/K. If, however, we view f as defined only over $U \cap V$, then $f^{-1}(K) = [u\cap(U\cap V)](U\cap v) = (u\cap V)(U\cap v)$ so that $(U\cap V)/(u\cap V)(U\cap v)$ is also isomorphic to H/K.

Thus the first and third of the above factor groups are isomorphic to each other. Similarly, the second and third factor groups are isomorphic.

Corollary 1. If H is a subgroup and N a normal subgroup of the group G, then $H/H \cap N$ is isomorphic to HN/N, a subgroup of G/N.

Proof: Set $G = U$, $N = u$, $H = V$ and the identity $1 = v$ in Theorem 1.

Corollary 2. Under the conditions of Corollary 1, if G/N is abelian, so also is $H/H \cap N$.

Let us call a group G solvable if it contains a sequence of subgroups $G = G_o \supset G_1 \supset \ldots \supset G_s = 1$, each a normal subgroup of the preceding, and with G_{i-1}/G_i abelian.

THEOREM 2. Any subgroup of a solvable group is solvable. For let H be a subgroup of G, and call $H_i = H \cap G_i$. Then that H_{i-1}/H_i is abelian follows from Corollary 2 above, where G_{i-1}, G_i and H_{i-1} play the role of G, N and H.

THEOREM 3. The homomorph of a solvable group is solvable.

Let $f(G) = G'$, and define $G'_i = f(G_i)$ where G_i belongs to a a sequence exhibiting the solvability of G. Then by (c) there exists a homomorphism mapping G_{i-1}/G_i on G'_{i-1}/G'_i. But the homomorphic image of an abelian group is abelian so that the groups G'_i exhibit the solvability of G'.

B. Permutation Groups.

Any one to one mapping of a set of n objects on itself is called a permutation. The iteration of two such mapping is called their product.

It may be readily verified that the set of all such mappings forms a group in which the unit is the identity map. The group is called the symmetric group on n letters.

Let us for simplicity denote the set of n objects by the numbers $1, 2, \ldots, n$. The mapping S such that $S(i) = i + 1 \bmod n$ will be denoted by $(123\ldots n)$ and more generally $(ij\ldots m)$ will denote the mapping T such that $T(i) = j, \ldots, T(m) = i$. If $(ij\ldots m)$ has k numbers, then it will be called a k cycle. It is clear that if $T = (ij\ldots s)$ then $T^{-1} = (s\ldots ji)$.

We now establish the

Lemma. If a subgroup U of the symmetric group on n letters $(n > 4)$ contains every 3-cycle, and if u is a normal subgroup of U such that U/u is abelian, then u contains every 3-cycle.

Proof: Let f be the natural homomorphism $f(U) = U/u$ and let $x = (ijk)$, $y = (krs)$ be two elements of U, where i, j, k, r, s are 5 numbers. Then since U/u is abelian, setting $f(x) = x'$, $f(y) = y'$ we have $f(x^{-1}y^{-1}xy) = x'^{-1}y'^{-1}x'y' = 1$, so that $x^{-1}y^{-1}xy \in u$. But $x^{-1}y^{-1}xy = (kji)\cdot(srk)\cdot(ijk)\cdot(krs) = (kjs)$ and for each k, j, s we have $(kjs) \in u$.

THEOREM 4. The symmetric group G on n letters is not solvable for $n > 4$.

If there were a sequence exhibiting the solvability, since G contains every 3-cycle, so would each succeeding group, and the sequence could not end with the unit.

C. Solution of Equations by Radicals.

The extension field E over F is called an extension by radicals if there exist intermediate fields $B_1, B_2, \ldots, B_r = E$ and $B_i = B_{i-1}(\alpha_i)$ where each α_i is a root of an equation of the form $x^{n_i} - a_i = 0$, $a_i \in B_{i-1}$. A polynomial f(x) in a field F is said to be solvable by radicals if its splitting field lies in an extension by radicals. We assume unless otherwise specified that the base field has characteristic 0 and that F contains as many roots of unity as are needed to make our subsequent statements valid.

Let us remark first that any extension of F by radicals can always be extended to an extension of F by radicals which is normal over F. Indeed B_1 is a normal extension of B_o since it contains not only α_1, but $\epsilon\alpha_1$, where ϵ is any n_1-root of unity, so that B_1 is the splitting field of $x^{n_1} - a_1$. If $f_1(x) = \prod_\sigma (x^{n_2} - \sigma(a_2))$, where σ takes all values in the group of automorphisms of B_1 over B_o, then f_1 is in B_o, and adjoining successively the roots of $x^{n_2} - \sigma(a_2)$ brings us to an extension of B_2 which is normal over F. Continuing in this way we arrive at an extension of E by radicals which will be normal over F. We now prove

THEOREM 5. The polynomial f(x) is solvable by radicals if and only if its group is solvable.

Suppose f(x) is solvable by radicals. Let E be a normal extension of F by radicals containing the splitting field B of f(x), and call G the group of E over F. Since for each i, B_i is a Kummer extension of B_{i-1}, the group of B_i over B_{i-1} is abelian. In the sequence of groups

$G = G_{B_o} \supset G_{B_1} \supset \ldots \supset G_{B_r} = 1$ each is a normal subgroup of the preceding since $G_{B_{i-1}}$ is the group of E over B_{i-1} and B_i is a normal extension of B_{i-1}. But $G_{B_{i-1}}/G_{B_i}$ is the group of B_i over B_{i-1} and hence is abelian. Thus G is solvable. However, G_B is a normal subgroup of G, and G/G_B is the group of B over F, and is therefore the group of the polynomial f(x). But G/G_B is a homomorph of the solvable group G and hence is itself solvable.

On the other hand, suppose the group G of f(x) to be solvable and let E be the splitting field. Let $G = G_o \supset G_1 \supset \ldots \supset G_r = 1$ be a sequence with abelian factor groups. Call B_i the fixed field for G_i. Since G_{i-1} is the group of E over B_{i-1} and G_i is a normal subgroup of G_{i-1}, then B_i is normal over B_{i-1} and the group G_{i-1}/G_i is abelian. Thus B_i is a Kummer extension of B_{i-1}, hence is splitting field of a polynomial of the form $(x^n-a_1)(x^n-a_2)\ldots(x^n-a_s)$ so that by forming the successive splitting fields of the $x^n - a_k$ we see that B_i is an extension of B_{i-1} by radicals, from which it follows that E is an extension by radicals.

Remark. The assumption that F contains roots of unity is not necessary in the above theorem. For if f(x) has a solvable group G, then we may adjoin to F a primitive n^{th} root of unity, where n is, say, equal to the order of G. The group of f(x) when considered as lying in F' is, by the theorem on Natural Irrationalities, a subgroup G' of G, and hence is solvable. Thus the splitting field over F' of f(x) can be obtained by radicals. Conversely, if the splitting field E over F of f(x) can be obtained by radicals, then by adjoining a suitable root of unity E is extended to E' which is still normal over F'. But E' could be obtained by adjoining first the root of unity, and then the radicals, to F; F would first be extended to F' and then F' would be extended to E'. Calling G the group of E' over F and G' the group of E' over F', we see that G' is solvable and G/G' is the group of F' over F and hence abelian. Thus G is solvable. The factor group G/G_E is the group of f(x) and being a homomorph of a solvable group is also solvable.

D. The General Equation of Degree n.

If F is a field, the collection of rational expressions in the variables $u_1, u_2, \ldots, u_n$ with coefficients in F is a field $F(u_1, u_2, \ldots, u_n)$. By the general equation of degree n we mean the equation

(1) $$f(x) = x^n - u_1x^{n-1} + u_2x^{n-2} - + \ldots + (-1)^n u_n.$$

Let E be the splitting field of f(x) over $F(u_1, u_2, \ldots, u_n)$. If $v_1, v_2, \ldots, v_n$ are the roots of f(x) in E, then

$u_1 = v_1 + v_2 + \ldots + v_n$, $u_2 = v_1 v_2 + v_1 v_3 + \ldots + v_{n-1} v_n, \ldots$
$\ldots, u_n = v_1 \cdot v_2 \cdot \ldots \cdot v_n$.

We shall prove that the group of E over $F(u_1, u_2, \ldots, u_n)$ is the symmetric group.

Let $F(x_1, x_2, \ldots, x_n)$ be the field generated from F by the variables $x_1, x_2, \ldots, x_n$. Let $\alpha_1 = x_1 + x_2 + \ldots + x_n$, $\alpha_2 = x_1x_2 + x_1x_3 + \ldots + x_{n-1}x_n, \ldots, \alpha_n = x_1x_2 \ldots x_n$ be the elementary symmetric functions, i.e., $(x-x_1)(x-x_2)\ldots(x-x_n) = x^n - \alpha_1 x^{n-1} + - \ldots (-1)^n \alpha_n = f^*(x)$. If $g(\alpha_1, \alpha_2, \ldots, \alpha_n)$ is a polynomial in $\alpha_1, \ldots, \alpha_n$, then $g(\alpha_1, \alpha_2, \ldots, \alpha_n) = 0$ only if g is the

zero polynomial. For if $g(\Sigma x_i, \Sigma x_i x_k, \ldots) = 0$, then this relation would hold also if the x_i were replaced by the v_i. Thus,
$g(\Sigma v_i, \Sigma v_i v_k, \ldots) = 0$ or $g(u_1, u_2, \ldots, u_n) = 0$ from which it follows that g is identically zero.

Between the subfield $F(\alpha_1, \ldots, \alpha_n)$ of $F(x_1, \ldots, x_n)$ and $F(u_1, u_2, \ldots, u_n)$ we set up the following correspondence: Let $f(u_1, \ldots, u_n)/g(u_1, \ldots, u_n)$ be an element of $F(u_1, \ldots, u_n)$. We make this correspond to $f(\alpha_1, \ldots, \alpha_n)/g(\alpha_1, \ldots, \alpha_n)$. This is clearly a mapping of $F(u_1, u_2, \ldots, u_n)$ on all of $F(\alpha_1, \ldots, \alpha_n)$. Moreover, if
$f(\alpha_1, \alpha_2, \ldots, \alpha_n)/g(\alpha_1, \alpha_2, \ldots, \alpha_n)$
$= f_1(\alpha_1, \alpha_2, \ldots, \alpha_n)/g_1(\alpha_1, \alpha_2, \ldots, \alpha_n)$, then $fg_1 - gf_1 = 0$. But this implies by the above that
$f(u_1, \ldots, u_n) \cdot g_1(u_1, \ldots, u_n) - g(u_1, \ldots, u_n) \cdot f_1(u_1, \ldots, u_n) = 0$
so that $f(u_1, \ldots, u_n)/g(u_1, u_2, \ldots, u_n)$
$= f_1(u_1, \ldots, u_n)/g_1(u_1, u_2, \ldots, u_n)$. It follows readily from this that the mapping of $F(u_1, u_2, \ldots, u_n)$ on $F(\alpha_1, \alpha_2, \ldots, \alpha_n)$ is an isomorphism. But under this correspondence $f(x)$ corresponds to $f^*(x)$. Since E and $F(x_1, x_2, \ldots, x_n)$ are respectively splitting fields of $f(x)$ and $f^*(x)$, by Theorem 10 the isomorphism can be extended to an isomorphism between E and $F(x_1, x_2, \ldots, x_n)$. Therefore, the group of E over $F(u_1, u_2, \ldots, u_n)$ is isomorphic to the group of $F(x_1, x_2, \ldots, x_n)$ over $F(\alpha_1, \alpha_2, \ldots, \alpha_n)$.

Each permutation of $x_1, x_2, \ldots, x_n$ leaves $\alpha_1, \alpha_2, \ldots, \alpha_n$ fixed and, therefore, induces an automorphism of $F(x_1, x_2, \ldots, x_n)$ which leaves $F(\alpha_1, \alpha_2, \ldots, \alpha_n)$ fixed. Conversely, each automorphism of $F(x_1, x_2, \ldots, x_n)$ which leaves $F(\alpha_1, \ldots, \alpha_n)$ fixed must permute the roots $x_1, x_2, \ldots, x_n$ of $f^*(x)$ and is completely determined by the permutation it effects on $x_1, x_2, \ldots, x_n$. Thus, the group of $F(x_1, x_2, \ldots, x_n)$ over $F(\alpha_1, \alpha_2, \ldots, \alpha_n)$ is the symmetric group on n letters. Because of the isomorphism between $F(x_1, \ldots, x_n)$ and E, the group for E over $F(u_1, u_2, \ldots, u_n)$ is also the symmetric group. If we remark that the symmetric group for $n > 4$ is not solvable, we obtain from the theorem on solvability of equations the famous theorem of Abel:

THEOREM 6. The group of the general equation of degree n is the symmetric group on n letters. The general equation of degree n is not solvable by radicals if $n > 4$.

E. Solvable Equations of Prime Degree.

The group of an equation can always be considered as a permutation group. If $f(x)$ is a polynomial in a field F, let $\alpha_1, \alpha_2, \ldots, \alpha_n$ be the roots of $f(x)$ in the splitting field $E = F(\alpha_1, \ldots, \alpha_n)$. Then each automorphism of E over F maps each root of $f(x)$ into a root of $f(x)$, that is, permutes the roots. Since E is generated by the roots of $f(x)$, different automorphisms must effect distinct permutations. Thus, the group of E over F is a permutation group acting on the roots $\alpha_1, \alpha_2, \ldots, \alpha_n$ of $f(x)$.

For an irreducible equation this group is always transitive. For let α and α' be any two roots of $f(x)$, where $f(x)$ is assumed irreducible. $F(\alpha)$ and $F(\alpha')$ are isomorphic where the isomorphism is the identity on F, and this isomorphism can be extended to an automorphism of E (Theorem 10). Thus, there is an automorphism sending any given root into any other root, which establishes the "transitivity" of the group.

A permutation σ of the numbers $1, 2, \ldots, q$ is called a linear substitution modulo q if there exists a number $b \not\equiv 0$ modulo q such that $\sigma(i) \equiv bi + c \pmod{q}$, $i = 1, 2, \ldots, q$.

THEOREM 7. Let $f(x)$ be an irreducible equation of prime degree q in a field F. The group G of $f(x)$ (which is a permutation group of the roots, or the numbers $1, 2, \ldots, q$) is solvable if and only if, after a suitable change in the numbering of the roots, G is a group of linear substitutions modulo q, and in the group G all the substitutions with $b = 1$, $\sigma(i) \equiv c + 1$ $(c = 1, 2, \ldots, q)$ occur.

Let G be a transitive substitution group on the numbers $1, 2, \ldots, q$ and let G_1 be a normal subgroup of G. Let $1, 2, \ldots, k$ be the images of 1 under the permutations of G_1; we say: $1, 2, \ldots, k$ is a domain of transitivity of G_1. If $i \leq q$ is a number not belonging to this domain of transitivity, there is a $\sigma \in G$ which maps 1 on i. Then $\sigma(1, 2, \ldots, k)$ is a domain of transitivity of $\sigma G_1 \sigma^{-1}$. Since G_1 is a normal subgroup of G, we have $G_1 = \sigma G_1 \sigma^{-1}$. Thus, $\sigma(1, 2, \ldots, k)$ is again a domain of transitivity of G_1 which contains the integer i and has k elements. Since i was arbitrary, the domains of transitivity of G_1 all contain k elements. Thus, the numbers $1, 2, \ldots, q$ are divided into a collection of mutually exclusive sets, each containing k elements, so that k is a divisor of q. Thus, in case q is a prime, either $k = 1$ (and then G_1 consists of the unit alone) or $k = q$ and G_1 is also transitive.

To prove the theorem, we consider the case in which G is solvable. Let $G = G_o \supset G_1 \supset \ldots \supset G_{s+1} = 1$ be a sequence exhibiting the solvability. Since G_s is abelian, choosing a cyclic subgroup of it would permit us to assume the term before the last to be cyclic, i.e., G_s is cyclic. If σ is a generator of G_s, σ must consist of a cycle containing all q of the numbers $1, 2, \ldots, q$ since in any other case G_s would not be transitive [if $\sigma = (1ij \ldots m)(n \ldots p) \ldots$ then the powers of σ would map 1 only into $1, i, j \ldots m$, contradicting the transitivity of G_s]. By a change in the number of the permutation letters, we may assume

$$\sigma(i) \equiv i + 1 \pmod q$$
$$\sigma^c(i) \equiv i + c \pmod q$$

Now let τ be any element of G_{s-1}. Since G_s is a normal subgroup of G_{s-1}, $\tau\sigma\tau^{-1}$ is an element of G_s, say $\tau\sigma\tau^{-1} = \sigma^b$. Let $\tau(i) = j$ or $\tau^{-1}(j) = i$ then $\tau\sigma\tau^{-1}(j) = \sigma^b(j) \equiv j + b \pmod q$. Therefore, $\tau\sigma(i) \equiv \tau(i) + b \pmod q$ or $\tau(i+1) \equiv \tau(i) + b$ for each i. Thus, setting $\tau(0) = c$, we have $\tau(1) \equiv c + b$, $\tau(2) \equiv \tau(1) + b = c + 2b$ and in general $\tau(i) \equiv c + ib \pmod q$. Thus, each substitution in G_{s-1} is a linear substitution. Moreover, the only elements of G_{s-1} which leave no element fixed belong to G_s, since for each $a \not\equiv 1$, there is an i such that $ai + b \equiv i \pmod q$ [take i such that $(a-1)i \equiv -b$].

We prove by an induction that the elements of G are all linear substitutions, and that the only cycles of q letters belong to G_s. Suppose the assertion true of G_{s-n}. Let $\tau \in G_{s-n-1}$ and let σ be a cycle which belongs to G_s (hence also to G_{s-n}). Since the transform of a cycle is a cycle, $\tau^{-1}\sigma\tau$ is a cycle in G_{s-n} and hence belongs to G_n. Thus $\tau^{-1}\sigma\tau = \sigma^b$ for some b. By the argument in the preceding paragraph, τ is a linear substitution $bi + c$ and if τ itself does not belong to G_s, then τ leaves one integer fixed and hence is not a cycle of q elements.

We now prove the second half of the theorem. Suppose G is a group of linear substitutions which contains a subgroup N of the form $\sigma(i) \equiv i + c$. Since the only linear substitutions which do not leave an integer fixed belong to N, and since the transform of a cycle of q elements is again a cycle of q elements, N is a normal subgroup of G. In each coset $N \cdot \tau$ where $\tau(i) \equiv bi + c$ the substitution $\sigma^{-1}\tau$ occurs, where $\sigma \equiv i + c$. But $\sigma^{-1}\tau(i) \equiv (bi + c) - c \equiv bi$. Moreover, if $\tau(i) \equiv bi$ and $\tau'(i) \equiv b'i$ then $\tau\tau'(i) \equiv bb'i$. Thus, the factor group (G/N) is isomorphic to a multiplicative subgroup of the numbers $1, 2, \ldots, q-1 \bmod q$ and is therefore abelian. Since (G/N) and N are both abelian, G is solvable.

<u>Corollary 1.</u> <u>If G is a solvable transitive substitution group on q letters (q prime), then the only substitution of G which leaves two or more letters fixed is the identity.</u>

This follows from the fact that each substitution is linear modulo q and $bi + c \equiv i \pmod q$ has either no solution $(b \equiv 1,\ c \not\equiv 0)$ or exactly one solution $(b \not\equiv 1)$ unless $b \equiv 1,\ c \equiv 0$ in which case the substitution is the identity.

<u>Corollary 2.</u> <u>A solvable, irreducible equation of prime degree in a field which is a subset of the real numbers has either one real root or all its roots are real.</u>

The group of the equation is a solvable transitive substitution group on q (prime) letters. In the splitting field (contained in the field of complex numbers) the automorphism which maps a number into its complex conjugate would leave fixed all the real numbers. By Corollary

1, if two roots are left fixed, then all the roots are left fixed, so that if the equation has two real roots all its roots are real.

F. Ruler and Compass Constructions.

Suppose there is given in the plane a finite number of elementary geometric figures, that is, points, straight lines and circles. We seek to construct others which satisfy certain conditions in terms of the given figures.

Permissible steps in the construction will entail the choice of an arbitrary point interior to a given region, drawing a line through two points and a circle with given center and radius, and finally intersecting pairs of lines, or circles, or a line and circle.

Since a straight line, or a line segment, or a circle is determined by two points, we can consider ruler and compass constructions as constructions of points from given points, subject to certain conditions.

If we are given two points we may join them by a line, erect a perpendicular to this line at, say, one of the points and, taking the distance between the two points to be the unit, we can with the compass lay off any integer n on each of the lines. Moreover, by the usual method, we can draw parallels and can construct m/n. Using the two lines as axes of a cartesian coordinate system, we can with ruler and compass construct all points with rational coordinates.

If a, b, c, . . . are numbers involved as coordinates of points which determine the figures given, then the sum, product, difference and quotient of any two of these numbers can be constructed. Thus, each element of the field $R(a, b, c, \ldots)$ which they generate out of the rational numbers can be constructed.

It is required that an arbitrary point is any point of a given region. If a construction by ruler and compass is possible, we can always choose our arbitrary points as points having rational coordinates. If we join two points with coefficients in $R(a, b, c, \ldots)$ by a line, its equation will have coefficients in $R(a, b, c, \ldots)$ and the intersection of two such lines will be a point with coordinates in $R(a, b, c, \ldots)$. The equation of a circle will have coefficients in the field if the circle passes through three points whose coordinates are in the field or if its center and one point have coordinates in the field. However, the coordinates of the intersection of two such circles, or a straight line and circle, will involve square roots.

It follows that if a point can be constructed with a ruler and compass, its coordinates must be obtainable from $R(a, b, c, \ldots)$ by a formula only involving square roots, that is, its coordinates will lie in a field $R_s \supset R_{s-1} \supset \ldots \supset R_1 = R(a, b, c, \ldots)$ where each field R_i is splitting field over R_{i-1} of a quadratic equation $x^2 - \alpha = 0$. It follows (Theorem 6, p. 21) since either $R_i = R_{i-1}$ or $(R_i/R_{i-1}) = 2$, that (R_s/R_1) is a power of two. If x is the coordinate of a constructed point, then $(R_1(x)/R_1)\cdot(R_s/R_1(x)) = (R_s/R_1) = 2^\nu$ so that $R_1(x)/R_1$ must also be a power of two.

Conversely, if the coordinates of a point can be obtained from $R(a, b, c, \ldots)$ by a formula involving square roots only, then the point can be constructed by ruler and compass. For, the field operations of addition, subtraction, multiplication and division may be performed by ruler and compass constructions and, also, square roots using $1 : r = r : r_1$ to obtain $r = \sqrt{r_1}$ may be performed by means of ruler and compass instructions.

As an illustration of these considerations, let us show that it is impossible to trisect an angle of 60°. Suppose we have drawn the unit circle with center at the vertex of the angle, and set up our coordinate system with X-axis as a side of the angle and origin at the vertex.

Trisection of the angle would be equivalent to the construction of the point (cos 20°, sin 20°) on the unit circle. From the equation $\cos 3\theta = 4\cos^3\theta - 3\cos\theta$, the abscissa would satisfy $4x^3 - 3x = 1/2$. The reader may readily verify that this equation has no rational roots, and is therefore irreducible in the field of rational numbers. But since we may assume only a straight line and unit length given, and since the 60° angle can be constructed, we may take $R(a, b, c, \ldots)$ to be the field R of rational numbers. A root α of the irreducible equation $8x^3 - 6x - 1 = 0$ is such that $(R(\alpha)/R) = 3$, and not a power of two.

A CATALOG OF SELECTED

DOVER BOOKS

IN SCIENCE AND MATHEMATICS

Math–Geometry and Topology

ELEMENTARY CONCEPTS OF TOPOLOGY, Paul Alexandroff. Elegant, intuitive approach to topology from set-theoretic topology to Betti groups; how concepts of topology are useful in math and physics. 25 figures. 57pp. 5⅜ x 8½. 60747-X

COMBINATORIAL TOPOLOGY, P. S. Alexandrov. Clearly written, well-organized, three-part text begins by dealing with certain classic problems without using the formal techniques of homology theory and advances to the central concept, the Betti groups. Numerous detailed examples. 654pp. 5⅜ x 8½. 40179-0

EXPERIMENTS IN TOPOLOGY, Stephen Barr. Classic, lively explanation of one of the byways of mathematics. Klein bottles, Moebius strips, projective planes, map coloring, problem of the Koenigsberg bridges, much more, described with clarity and wit. 43 figures. 210pp. 5⅜ x 8½. 25933-1

CONFORMAL MAPPING ON RIEMANN SURFACES, Harvey Cohn. Lucid, insightful book presents ideal coverage of subject. 334 exercises make book perfect for self-study. 55 figures. 352pp. 5⅝ x 8¼. 64025-6

THE GEOMETRY OF RENÉ DESCARTES, René Descartes. The great work founded analytical geometry. Original French text, Descartes's own diagrams, together with definitive Smith-Latham translation. 244pp. 5⅜ x 8½. 60068-8

THE THIRTEEN BOOKS OF EUCLID'S ELEMENTS, translated with introduction and commentary by Sir Thomas L. Heath. Definitive edition. Textual and linguistic notes, mathematical analysis. 2,500 years of critical commentary. Unabridged. 1,414pp. 5⅜ x 8½. Three-vol. set.
Vol. I: 60088-2 Vol. II: 60089-0 Vol. III: 60090-4

GEOMETRY OF COMPLEX NUMBERS, Hans Schwerdtfeger. Illuminating, widely praised book on analytic geometry of circles, the Moebius transformation, and two-dimensional non-Euclidean geometries. 200pp. 5⅝ x 8¼. 63830-8

DIFFERENTIAL GEOMETRY, Heinrich W. Guggenheimer. Local differential geometry as an application of advanced calculus and linear algebra. Curvature, transformation groups, surfaces, more. Exercises. 62 figures. 378pp. 5⅜ x 8½. 63433-7

CURVATURE AND HOMOLOGY: Enlarged Edition, Samuel I. Goldberg. Revised edition examines topology of differentiable manifolds; curvature, homology of Riemannian manifolds; compact Lie groups; complex manifolds; curvature, homology of Kaehler manifolds. New Preface. Four new appendixes. 416pp. 5⅜ x 8½. 40207-X

TOPOLOGY, John G. Hocking and Gail S. Young. Superb one-year course in classical topology. Topological spaces and functions, point-set topology, much more. Examples and problems. Bibliography. Index. 384pp. 5⅝ x 8¼. 65676-4

Physics

OPTICAL RESONANCE AND TWO-LEVEL ATOMS, L. Allen and J. H. Eberly. Clear, comprehensive introduction to basic principles behind all quantum optical resonance phenomena. 53 illustrations. Preface. Index. 256pp. 5⅜ x 8½. 65533-4

ULTRASONIC ABSORPTION: An Introduction to the Theory of Sound Absorption and Dispersion in Gases, Liquids and Solids, A. B. Bhatia. Standard reference in the field provides a clear, systematically organized introductory review of fundamental concepts for advanced graduate students, research workers. Numerous diagrams. Bibliography. 440pp. 5⅜ x 8½. 64917-2

QUANTUM THEORY, David Bohm. This advanced undergraduate-level text presents the quantum theory in terms of qualitative and imaginative concepts, followed by specific applications worked out in mathematical detail. Preface. Index. 655pp. 5⅜ x 8½. 65969-0

ATOMIC PHYSICS (8th edition), Max Born. Nobel laureate's lucid treatment of kinetic theory of gases, elementary particles, nuclear atom, wave-corpuscles, atomic structure and spectral lines, much more. Over 40 appendices, bibliography. 495pp. 5⅜ x 8½. 65984-4

AN INTRODUCTION TO HAMILTONIAN OPTICS, H. A. Buchdahl. Detailed account of the Hamiltonian treatment of aberration theory in geometrical optics. Many classes of optical systems defined in terms of the symmetries they possess. Problems with detailed solutions. 1970 edition. xv + 360pp. 5⅜ x 8½. 67597-1

THIRTY YEARS THAT SHOOK PHYSICS: The Story of Quantum Theory, George Gamow. Lucid, accessible introduction to influential theory of energy and matter. Careful explanations of Dirac's anti-particles, Bohr's model of the atom, much more. 12 plates. Numerous drawings. 240pp. 5⅜ x 8½. 24895-X

ELECTRONIC STRUCTURE AND THE PROPERTIES OF SOLIDS: The Physics of the Chemical Bond, Walter A. Harrison. Innovative text offers basic understanding of the electronic structure of covalent and ionic solids, simple metals, transition metals and their compounds. Problems. 1980 edition. 582pp. 6⅛ x 9¼. 66021-4

HYDRODYNAMIC AND HYDROMAGNETIC STABILITY, S. Chandrasekhar. Lucid examination of the Rayleigh-Benard problem; clear coverage of the theory of instabilities causing convection. 704pp. 5⅜ x 8¼. 64071-X

INVESTIGATIONS ON THE THEORY OF THE BROWNIAN MOVEMENT, Albert Einstein. Five papers (1905–8) investigating dynamics of Brownian motion and evolving elementary theory. Notes by R. Fürth. 122pp. 5⅜ x 8½. 60304-0

THE PHYSICS OF WAVES, William C. Elmore and Mark A. Heald. Unique overview of classical wave theory. Acoustics, optics, electromagnetic radiation, more. Ideal as classroom text or for self-study. Problems. 477pp. 5⅜ x 8½. 64926-1